Fluorescence Imaging - Recent Advances and Applications

Edited by Raffaello Papadakis

Published in London, United Kingdom

Fluorescence Imaging - Recent Advances and Applications
http://dx.doi.org/10.5772/intechopen.100985
Edited by Raffaello Papadakis

Contributors
Mohamed Yahya, Müjgan Yaman, Zeynel Seferoğlu, Raffaello Papadakis, Hu Li, Yahui Li, Lixin Liu, Jinshou Tian, Dong Xiao, Natakorn Sapermsap, Yu Chen, Hang Li, David Day-Uei Li, Francisco Rodríguez-Vidal, Peter R. Corridon

First published in London, United Kingdom, 2023 by IntechOpen
IntechOpen is the global imprint of INTECHOPEN LIMITED, registered in England and Wales, registration number: 11086078, 5 Princes Gate Court, London, SW7 2QJ, United Kingdom

British Library Cataloguing-in-Publication Data
A catalogue record for this book is available from the British Library

Additional hard and PDF copies can be obtained from orders@intechopen.com

Fluorescence Imaging - Recent Advances and Applications
Edited by Raffaello Papadakis
p. cm.
Print ISBN 978-1-80355-183-8
Online ISBN 978-1-80355-184-5
eBook (PDF) ISBN 978-1-80355-185-2

For EU product safety concerns:
IN TECH d.o.o., Prolaz Marije Krucifikse Kozulić 3, 51000 Rijeka, Croatia,
info@intechopen.com or visit our website at intechopen.com.

Meet the editor

Dr. Raffaello Papadakis is a chemical engineer (MEng) with a Ph.D. in Physical Organic Chemistry from the Technical University of Athens, Greece. He has conducted research at various research institutes including CNRS/Aix-Marseille University, France and Ångström Laboratories, and Uppsala University, Sweden. His research interests revolve around physical organic chemistry, materials science, and photochemistry with a particular focus on the chemistry of graphene and polysaccharides as well as the development of novel fluorophores and bioconjugation. Dr. Papadakis has authored/co-authored more than forty peer-reviewed scientific papers and book chapters. Currently, he works as a senior research scientist at TdB Labs in Uppsala, Sweden, where he is particularly interested in the fluorescent labeling of polysaccharides.

Contents

Preface

Fluorescence imaging is a non-invasive technique that is widely known to enable visualization of biological processes occurring within living organisms. It encompasses various methods such as microscopy and spectroscopy and can involve a wide variety of fluorescent imaging probes.

In bioscience, fluorescence imaging constitutes an indispensable tool, which has over the last years revolutionized the field by empowering monitoring of cellular processes. By using fluorescent dyes or proteins, scientists can label specific molecules or structures within cells and track their behavior in real-time. Consequently, fluorescence imaging has enabled breakthroughs in understanding cellular functions, such as protein interactions, gene expression, and intracellular signaling pathways.

The use of fluorescence imaging is quite versatile and is not limited only to biological studies. During the last few years, it has also found applicability in a variety of research and technology areas including medical diagnostics, drug discovery, environmental monitoring, and materials science. For instance, fluorescence imaging can be used to detect cancer cells in tissue samples, monitor the distribution of drugs within the body, or assist in detecting microplastics and their degradation. Additionally, it has been highly contributory in studying the effects of pollutants on living organisms as well as for the characterization of nanomaterials.

Acknowledging the quick progress in the field, this book focuses on some new developments within fluorescence imaging. It is organized into three sections. The first section covers novel applications of fluorescence imaging, such as renal intravital microscopy and excitation-emission matrix fluorescence environmental applications. The second section pertains to the development of novel fluorophores and optical chemosensors with an emphasis on graphene-based materials and other optical chemosensors. Finally, the last section deals with the advancement of novel fluorescence imaging techniques and methodologies focusing on time-domain fluorescence lifetime imaging.

In my role as the academic editor of this publication, I would like to express my appreciation to all the authors who have contributed chapters encompassing various research and technology domains related to fluorescence imaging. It has been an honor to explore the groundbreaking works in this significant area of research. A big "thanks" goes to all of you.

I would also like to acknowledge TdB Labs AB, the company I have had the privilege to work for as a Senior Research Scientist since 2019. I would especially like to thank

Charlotte de Belder Tesséus (CEO of TdB Labs AB) for the enthusiasm and continuous encouragement she has provided as well as for her immense interest in fluorescent markers, which has been very motivating for me.

Dr. Raffaello Papadakis
TdB Labs AB,
Uppsala, Sweden

Section 1

Novel Applications of Fluorescence Imaging

Chapter 1

Fluorescent Dextran Applications in Renal Intravital Microscopy

Peter R. Corridon

Abstract

Dextrans, which is a generic term used to describe a family of glucans, are branched polysaccharide molecules derived from lactic acid bacteria in the presence of sucrose. These complex branched glucans have various uses in the medical industry, including plasma expanders and anticoagulants, and have also been investigated for their utility in targeted and sustained delivery of drugs, proteins, enzymes, and imaging agents for renal applications. Simultaneous advances in renal intravital microscopy have brought several advantages over in vitro and ex vivo models by providing real-time assessments of dynamic processes at the cellular and subcellular levels. Such advances have been used to support regenerative medicine strategies. Consequently, this chapter aims to provide an overview of how fluorescent dextrans have supported renal gene and cell therapies and evolving tissue engineering techniques.

Keywords: fluorescent, dextran, dextran sulphate, renal, intravital microscopy, two-photon microscopy, multiphoton microscopy, gene therapy, cell therapy, regenerative medicine

1. Introduction

Historically dextrans have been extensively used in clinical practice and experimental models. These compounds can be prepared in a wide range of molecular weights to support their application throughout the vasculature. For instance, intravascular infusions of relatively large molecular weight dextrans have been used as plasma expanders in hypovolemic treatments, substitutes for whole blood in cases of severe shock, and a mode to combat volume depletion in various conditions [1–7]. In contrast, low molecular weight dextrans, which can be filtered by the glomerulus and reclaimed by the tubules, have found use in nephropathology evaluations in experimental and clinical approaches [8–17]. Moreover, within other branches of the medical industry, these complex branched glucans possess properties that support anticoagulation and allow them to be effectively used as vehicles for the targeted and sustained delivery of drugs, proteins, enzymes, and imaging agents [6, 18].

Simultaneously, microscopy advances, particularly intravital microscopic techniques, have brought several advantages over in vitro and ex vivo models by providing instant assessments of dynamic cellular and subcellular events in vivo [13, 14, 19–23]. As we have also improved our capacities to induce genetic alterations within the past two decades effectively, new opportunities have arisen to, in turn, improve our fundamental understanding of mechanisms that drive disease progression and define novel gene- and cell-based strategies to combat these debilitating outcomes. To date, it can be argued that renal intravital microscopy (IVM) has benefited substantially from the intersection of these approaches. Consequently, this chapter can provide insight into how fluorescent dextrans have supported the development of renal gene and cell therapies and describe how these approaches are impacting the evolving field of regenerative medicine, specifically within the realm of whole organ bioengineering.

2. How optical microscopic techniques have advanced kidney imaging

2.1 Optical microscopy

Optical microscopy systems can be classified into two major categories: linear and non-linear. Traditionally, confocal fluorescence and wide-field laser microscopy systems have been regarded as the most prominent applications [24], and such linear techniques are extensively applied in in vitro studies focused on cell culture and tissue sections. One major challenge with these approaches revolves around spatial resolution. Spatial resolution is generally restricted and merely provides a viable imaging depth from the specimen's surface of roughly 100 μm. Compounded scattering events constrain light propagation and hinder image acquisition at greater depths [25]. Such characteristics limit in vivo applications and the overall clinical utility of these systems. Specifically, these matters are of particular significance to confocal microscopic applications, which rely on the light emanating from the sample that is, in turn, channeled to the pinhole. High degrees of scattering generated in turbid media are well-correlated with tissue depth in biological tissues [13], which also limits the intensity of light that can propagate through the pinhole, thereby severely attenuating the optical signals.

Non-linear microscopic approaches, by comparison, offer the use of higher-order light-matter interactions with multiple photons. This process inherently enhances image-based contrast and, as a result, considerably surmounts the obstacles hindering tissue imaging at greater depths [26]. This principle is widely applied in multiphoton (namely, two-photon) absorption fluorescence excitation techniques that utilize quadratic (non-linear) relationships between excitation and emission events [27, 28]. In doing do, two-photon systems facilitate investigations in tissues at imaging depths on the order of a millimeter, significantly exceeding conventional single-photon processes like confocal microscopy. These systems require the simultaneous absorption of two photons for this excitation process to generate emissions that vary with the square of the excitation intensity.

Multiphoton imaging thus provides several benefits over its counterparts, including reduced scattering levels with infrared-derived excitations, out-of-focus photobleaching, and background fluorescence produced by localized excitations [17]. Overall, these benefits have supported multiphoton imaging over other light-based microscopy systems for live biological tissue examinations, permitting deep tissue imaging at a high resolution. Furthermore, this advanced imaging technique aids the monitoring of live physiological/pathophysiological cellular and subcellular events in real-time. It is precisely for these reasons that multiphoton fluorescence microscopy has gained widespread application in the intravital imaging of the kidney. These applications include analyzing normal renal physiology [13, 28, 29] and pathophysiology [30], and specifically, the quantification of acute, chronic, and end-stage disorders [16, 31], molecular biodistributions, and effects of various drug development [32], monitoring renal genetic alterations, and most recently tissue engineering [33].

2.2 Intravital microscopy

Advanced imaging technologies, such as multiphoton microscopy, have given researchers powerful tools to answer critical problems in live systems. One such tool is IVM. This form of optical microscopy provides investigations at the cellular and subcellular levels [34]. The development of such non-linear approaches has resulted in a considerable surge of in vivo investigations to tackle fundamental issues in various organ systems [35].

IVM has demonstrated tremendous utility in studying innate and adapted morphology and functional processes among all forms of microscopy. IVM, in particular, has revolutionized our knowledge of living systems through detailed dynamic insights on various regulatory processes [36–40], as well as in-depth anatomical distinctions that support these processes. Non- or minimally-invasive multiphoton systems generate high contrast images with exquisite lateral spatial resolutions that can be adapted to acquire three-dimensional volumes in time-dependent manners [24]. These facts make this system well-suited for tissue, cellular, and molecular studies [41] and have developed new avenues for studying live systems in physiological and pathophysiological conditions, particularly within the kidney, through enhanced live four-dimension applications. Therefore, it is believed that this modality will continue to transform this industry as new ways are being sought after to improve image acquisition time and imaging depth and reduce the complexity and cost of these systems, Overall, these benefits can, someday, drive its clinical potential and utility in the field of biomedicine [41].

3. Fluorescent dextrans and intravital microscopy

3.1 The dextran

One major component that has helped advance this fields of study is the dextran. Dextrans are complex branched glucans produced from glucose polymers via chemical synthesis or from the bacterium Leuconostoc mesenteroides action on sucrose bacteria [18]. These polysaccharides are primarily made up of linear α-1,6-glucosidic linkages with various degrees of branching. Specifically, the glucose subunits are linked by $\alpha(1 \rightarrow 6)$ linkages on its main chain and $\alpha(1 \rightarrow 3)$ linkages on its side chain [42]. Their polymerization is catalyzed by the enzyme dextransucrase and can occur in several microorganisms to produce compounds that vary in molecular weight. The resulting dextrans can also vary in branching patterns, from slightly to highly branched structures.

The polydispersity of dextrans is another physicochemical feature that influences their behavior in vivo and their excellent biocompatibility. Dextran polymers are generally stable under normal, as well as mildly acidic and basic conditions. Dextrans also contain a significant number of hydroxyl groups for conjugation, supporting their high-water solubility. These heterogeneous polysaccharides are biodegradable and can be cleaved by various dextranases. Altogether these properties have supported their preclinical and clinical use for the past several decades.

3.2 Clinical- and research-based applications of dextran molecules

Dextrans produce relatively low-viscosity solutions and are generally classified as neutral polymers. Based on this classification, dextrans are versatile compounds with many clinical applications in anesthesiology, radiology, ophthalmology, and emergency medicine. Typically, these compounds cannot penetrate the intact membranes of living cells. As a result, they can be used to evaluate membrane dynamics, and macromolecular distributions and kinetics [43]. Specifically, large molecular weight dextrans in the plasma can be used to assess and mimic the properties of the circulation in both normal and pathological conditions [44].

Initially, these molecules were used as colloids for fluid resuscitation as they osmotically expand the plasma and help restore blood plasma volume post severe hemorrhage [45]. These compounds distribute throughout the circulation, expand blood volume, and thus, increase cardiac output and blood flow. However, these polymers appear to impede fibrin network formation by increasing this protein's degradation. Research has also shown that the significant presence of dextran molecules leads to decrements in von Willebrand factor altering platelet funcion [46]. As a result, a move that supported their general abandonment in such settings arose from the abovementioned issues, along with the adverse effects these compounds had on the innate coagulation system and their potential to induce anaphylactic responses [46]. Interestingly, this antithrombotic effect provides additional clinical utility for preventing postoperative venous thrombosis.

From a medical imaging perspective, these compounds can be labeled with various markers to support various non-invasive detection and diagnostic techniques. For instance, the intravenous delivery of dextrans labeled with technetium Tc-99 m serves as contrast agents for nuclear medicine, magnetic resonance imaging, or scintigraphy investigations [46]. Whereas, from a microscopic perspective, fluorescently-labeled dextrans have been widely used to support countless investigations within the kidney, which is the focus of this chapter. That is, dextrans are easily modified to accept fluorophores that can be used to label various renal compartments exclusively based on molecular weight. Specifically, this property facilitates the study of microvascular flow, vascular integrity, vesicular trafficking, glomerular filtration, and renal reabsorption and secretion [35]. Examples of some commonly used fluorescent dextrans are provided in **Figure 1**.

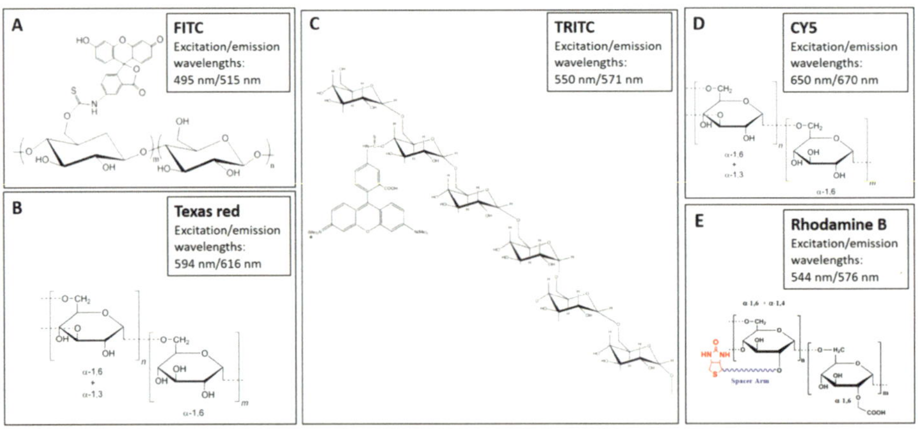

Figure 1.
The structures of some commonly used fluorescent dextrans. This image depicts the following fluorescently conjugated dextrans: (A) FITC, (B) Texas red, (C) TRITC, (D) cyanine 5 (CY5), and (E) rhodamine B.

3.3 Intravital multiphoton fluorescence microscopy fundamentals

The fields revolving around biological fluorescence have a rich and vibrate history that dates back to original observations presented in 1565 by Nicolas Bautista Monardes, botanist and physician [47]. Monardes' article communicated the simple outlines of various visible hues emanating from several wood types. However, such observations provided a means to detect counterfeit samples from scarce and valuable materials, like coatli, known for its diuretic properties. This simplified approach relied on a crucial characteristic of the counterfeit materials: their ability to emit clear blue hues after being immersed in water [48]. This characteristic was beneficial and undoubtedly paved a path for contemporary optical advancements.

As various fields in optics advanced, two noteworthy processes, incandescence and photoluminescence, have been defined. Incandescence describes a thermal radiative process that supports the generation of electromagnetic radiation from the thermal motion of charged particles within substances. The resulting electromagnetic radiation emanates from the visible spectrum and occurs as a consequence of elevations in temperature. Such a heat response generates increased particle motion, giving these particles a more remarkable ability to radiate. In comparison, photoluminescence also describes optical emissions via electronic state transitions, albeit in a manner that is independent of heat and is deemed a cold process. Due to this nature, this phenomenon was the center of much debate two centuries ago, when scientists vehemently argued about the applicability of photoluminescence within the realm of thermodynamics [48]. Nevertheless, we have witnessed the use and classification of several aspects of photoluminescence, namely, resonant radiation, phosphorescence, and fluorescence, which permeate our daily lives. Each of these three processes will be briefly outlined below.

We can first focus on the theory related to the swift emission of electromagnetic energy derived from the photonic absorption of gaseous atoms, upon which resonant radiation centers. Within this process, incident photons, generally, possess the same or perhaps frequencies similar to the resonant frequencies of the gaseous atoms, allowing them to first transition to higher energy levels after radiation absorption, and subsequently relax to a lower energy state. The latter transition is accompanied by the emission of photons with energy levels comparable to incident particles, eliminating substantial energy decrements during these transitions and defining discrete energy differentials. These discrete demarcations are characteristic of a given atom and thus provide unique energy transition signatures. Next, with phosphorescence, photonic emission results in a reduction of energy states. Compared to resonant radiation, in this excitation process, the emitted photons possess lower energy levels than their counterparts and generally occur on a longer timescale, allowing phosphorescent materials to discharge radiation for extended periods post excitation.

Building on the previously mentioned description provided on resonant radiation, we can first begin to examine one-photon or conventional fluorescence further. In this process, a single photon is absorbed by a fluorophore. Fluorophore atoms comprise electrons occupying various specific electronic states, defining their perpetual vibrational, rotational, or translational motions that occur in relation to the physical state of the fluorescing compound(s). Notably, the excitation of a fluorophore (or a luminophore) can occur through either single or multiphoton absorption events, and once a fluorophore absorbs a photon, a resulting electronic energy transition will occur. The probability that such a fluorophore will collide with surrounding molecules to support its relaxation is increased upon transitioning to a higher energy level.

Likewise, the de-excitation process can, in turn, be defined in three phases. Within the first phase, some of the absorbed energy can facilitate fluorophores' vibrational/rotational modes as well as heat generation from internal conversion-based radiative decay or radiationless de-excitation. The second phase commences and revolves around energy losses/conversion processes required to produce photons, as defined by the Stokes shift, which exists with energy levels lower than those of their incident counterparts. Ultimately, in this form of de-excitation, the third phase is characterized by an additional internal conversion process that further reduces the particles' energy states.

As a result, we can thus compare the processes mentioned above to the multiphoton phenomena utilized in IVM. With IVM, fluorescence is derived from the absorption of two or more low-energy photons concurrently. This combination is essential as each photon's energy is incapable of generating an excitation event, yet when combined, the resulting energy level is sufficient to facilitate this electronic transition. Moreover, as observed in single photon fluorescence, the de-excitations result in photonic emissions described in detail in the literature [28, 38, 39, 49].

3.4 Practical ways to generate multiphoton excitation fluorescence

In order to generate a multiphoton excitation event, the required photons must be absorbed by the fluorophore within a single attosecond. This constraint drastically minimizes the probability of naturally occurring multiphoton phenomena. To illustrate this point, if we expose a rhodamine molecule to direct sunlight, we can expect a one-photon excitatory event to occur within a timeframe of 1–2 seconds, whereas a two-photon event would occur after 10 million years [6, 11]. These depictions were revealed in 1931 by Dr. Marie Groppert-Mayer. She simultaneously forecasted that an enormous flux of incident radiation could overcome this 10 million-year period to produce detectable levels of multiphoton fluorescence in a time frame comparable to the one-photon excitatory event [28, 38, 39, 49].

Dr. Groppert-Mayer's models devised the underpinning framework upon which femtosecond laser technologies have been developed. Today, these robust systems, like titanium: sapphire lasers, produce high and sustained photon fluxes of energy required that facilitate routine multiphoton excitations/emissions from femtosecond pulses. The concise duration in which these infrared systems emit light pulses inherently only generates photon fluxes that can raise the temperature of water by, on average, 0.2 K/sec [28]. Furthermore, exposure to phototoxic effects is limited by the fact that long wavelength, low-energy photons are constricted to minimize scattering and improve depth penetration and resolution for safe and effective biological studies. Compared to the two-photon excitation, in single-photon excitation, tissues are generally exposed to far more levels of shorter wavelength and higher energy, ultraviolet and visible light with lasers that excite fluorophores within substantially greater volumes that subject tissues to more debilitating events. Excitingly, these benefits have propelled research to develop three-photon systems that require roughly 10-fold photon densities applied in two-photon microscopic applications and promise enhanced depth penetration and resolution [50].

3.5 Image formation in multiphoton fluorescence microscopy

The next step in the IVM process revolves around image acquisition. This process converts the optical signals that emanate from a sample to electrical signals. Extensive details of this process can be obtained in the literature [51], and for the purposes of this chapter, the fundamental concept is outlined as follows. First, it is essential to identify key components of a multiphoton fluorescence microscopy system that support image acquisition and formation. These components are the objectives, the mirrors, and the detectors, which are utilized in a subsequent manner.

After the mode-locked/pulse femtosecond laser emits incident radiation, scanning dichroic mirrors are used to guide the beams of light onto various objective lenses. These lenses then focus the beams at unitary loci within the sample. As theorized by Dr. Groppert-Mayer, this process directs enormous fluxes of incident radiation on the specimen in both spatially and temporally fashions and facilitates multiphoton fluorescence excitation events that emit photon beams. Despite applying enormous quantities of incident radiation, the system's average input power is delimited below 10 mW via low pulse duty cycles and coincides with the power levels generated by confocal systems. Also, since most photons follow a direct path, multiphoton systems have drastically enhanced signal-to-noise ratios compared to one-phase microscopy systems.

Further comparisons to confocal fluorescence microscopy identify the absence of a pinhole in IVM systems. The alternative configuration provides more flexibility in designating the detection geometry to incorporate descanned or non-descanned detection schemes. Non-descanned systems provide a means to enhance depth penetration and reduce the number of optical elements needed and the path length that emerging fluorescence signals can interact with dust particles, thereby limiting losses to scattering and improving the sensitivity of the system light and efficiency of light collection. This process enhances sensitivity without compromising image quality. It is essential for maximal depth penetration into living tissue, as the detectors contain susceptible photomultiplier tubes capable of detecting low levels of light and barrier filters that are used to generate red, green, and blue pseudo-color images formed from a 3-D geometrical correlation within a given specimen. Two-dimensional, XY-plane raster scanning processes support data acquisition that can be coupled with depth position (Z-plane). The resulting imaging datasets can be extended to 4-D investigations by acquiring data in a time-dependent manner.

3.6 In vivo multiphoton imaging of mammalian tissues and its benefits over ex vivo and in vitro approaches

Historically, researchers have relied on in vitro, ex vivo, and in vivo experimental models to answer scientific questions that have and continue to improve our mechanistic understanding of various physiological processes. Such work has spearheaded advancements that, for example, have allowed the examination of complex cellular-molecular interactions in vitro. In vitro characterizations provide opportunities to conduct high-throughput, cost-effective studies that are the first and integral initial step that circumvents complex intracellular, intercellular, intra-organ, and inter-organ events within live animal models. After that, such approaches can be transitioned to comparable ex vivo and, ultimately, in vivo models. Experimental approaches of this nature provide greater degrees of flexibility and facilitate ways to view complex phenomena more straightforwardly and systematically, and conduct practices that would be impractical for animal studies that may not be approved for animal usage by regulatory boards.

Nevertheless, in vitro and ex vivo models, by their very nature, are incapable of fully mimicking natural phenomena. The fact illustrates the trade-off that must be considered when solely relying on this form of experimentation and the need to apply multiple approaches when studying complex biological systems [52]. As a result, IVM has gained more prominence. These live imaging systems have equipped researchers with a means to acquire unique and compelling evidence that can only be gathered from whole organ investigations [9]. IVM's current utility can be extended through more invasive approaches that support organ exposure/exteriorization that can be conducted on rodent brains [53–55], livers [56–58], and kidneys [9, 58–60], as well as ex vivo applications that extend depth penetration of existing IVM systems. Besides these accomplishments, IVM is currently a niche platform that can only be performed non-invasively on shallow tissue depths and easily accessible organs like the skin [61–63].

The preparations to conduct such imaging studies are crucial in ensuring the generation of valuable microscopic data. In particular, intravital multiphoton imaging of kidney function and structure has become quite popular since the kidneys of rats and mice can be easily externalized after anesthesia and placed in the view of a microscope lens [12, 28, 35, 64, 65]. To conduct such studies, it is important to outline some of the following essential conditions that include the use of anesthesia, analgesics, and antiseptics, as well as other common surgical considerations that should be considered. Generally, standalone or combinative inhalable (isoflurane, for short duration studies) and injectable (for more extended duration studies: pentobarbital – for survival studies or thiobutabarbital – for terminal studies) sedatives are used for small animal studies. Analgesics like acetaminophen and antiseptics, like a surgical scrub, are also routinely used at the end of survival studies. Researchers should also compensate for fluid losses by introducing isotonic fluids and serum albumin to regulate osmotic pressures.

It is generally recommended to conduct surgical procedures in sterile environments, especially for survival studies. After fully sedating the rodent, one should constantly monitor its core temperature, often done with an anal probe, along with heating pads, lamps, and blankets to regulate core temperature during the surgical and imaging procedures. A carotid or femoral artery access catheter can also monitor blood pressure. Further safeguards may be taken by sterilizing the imaging dish and saline in which externalized kidneys are placed to limit infection and tissue dehydration/pH alterations.

4. Materials and methods for intravital studies

4.1 Animals and associated procedures for intravital studies

Primarily male Sprague Dawley rats (Harlan Laboratories, Indianapolis, IN), as well as Frömter Munich Wistar (Harlan Laboratories, Indianapolis, IN) and Simonsen Munich Wistar (Simonsen's Laboratory, Gilroy, CA) rats are used for these types of studies. These animals generally range in weight from 150 to 450 g. Wistar rats are used for glomerular studies due to the unique abundance of superficial glomeruli that can be easily accessed using this imaging technique. Additionally, all animals should be given free access to standard rat chow and water (unless the model requires otherwise, and most importantly, all experiments must be conducted under the approval of institutional animal care committees and welfare guidelines. These approaches can be adjusted for studies in other rodents, namely, mice. However, for this chapter, we will focus on rat models.

4.2 Fluorescent dextran marker preparation and infusions

Various conjugated dextrans, which vary according to the fluorescent tag and molecular weights, can be used for intravital two-photon fluorescent imaging studies. For example, single or combinations of the following dextrans can be applied to examine vascular integrity and routine renal filtration capacities: 3 kDa Cascade Blue, and 4 and 5 kDa Fluorescein Isothiocyanate (FITC) dextrans (Invitrogen, CA); and 150 kDa Tetramethyl Rhodamine Isothiocyanate (TRITC) dextran (TdB Consultancy, Uppsala, Sweden). It is recommended first to produce a stock solution that can be used in an imaging study with a concentration of 20 mg/ml, from which 500 µl can be diluted in 1 ml of isotonic saline [66]. Care should also be taken to ensure that the molecular weight of large dextran molecules should not be dispersed over a wide range, as such variations can support renal ultrafiltration and incorrectly report filtration and reabsorption capacities.

In live rats, two main modes of infusions are supported via jugular and tail veins (**Figure 2**). For jugular vein infusions, it is vital first to anesthetize the rat using isoflurane in 5% oxygen (Webster Veterinary Supply, Inc., Devens, MA). After this initial sedation, the animal's core temperature (approximately 37°C) should be regulated with a heating pad (**Figure 2**), and an intraperitoneal injection can be given for survival (50 mg/kg of pentobarbital) or non-survival (130 mg/kg thiobutabarbital) procedures. Once the rat is completely stabilized, the researcher can shave its neck and sanitize the region using a common antiseptic like Betadine Surgical Scrub (Purdue Products L.P., Stamford, CT). Incisions can then be made to expose and isolate the jugular vein with 3–0/4–0 silk loops. Common practice is applying a superior loop that can be tied and clamped with a pair of hemostats to stiffen and elevate this vein. After that, a minor incision can be made to facilitate the insertion of a PE-50 tubing catheter, which is attached to a 1 ml syringe containing injectate, into the jugular vein. Another silk loop can be applied to anchor the catheter further. Similarly, tail vein infusions can be conducted post sedation by moistening the tail with a warmed sheet of gauze or placing it into a warm bath to support dilation. Once dilated, A 25-gauge butterfly needle attached to a syringe containing injectates can be inserted into this vein to support the delivery of infusates.

Analogously, hydrodynamic retrograde renal vein fine-needle injections have been defined to facilitate renal cell, and gene transplantation [16, 17, 67]. In this process, intraperitoneal incisions are performed to isolate and occlude venous segments using delicate and non-traumatic micro-serrefine clamps of the left renal hilus. First renal artery is clamped, and then the renal vein. The vein is then elevated with either 3–0 or 4–0 silk loops to support rapid injections into this outport. The needle can then be removed, and pressure applied to the injection site using a cotton swab to induce hemostasis. Further details of this technique are outlined in the literature [16, 17, 67]. Lastly, the vascular clamps can be removed (the venous clamp should be removed before the arterial clamp) to restore flow. The total clamping should be less than 3 minutes and the midline incision can be closed to allow the animal to recover.

4.3 Renal intravital two-photon fluorescence microscopy

In anesthetized rats, the left flank should be shaved, and vertical incisions need to be created to externalize the left kidney. A heating pad can then be placed over the rat to maintain its core temperature. The investigator can then place the kidney inside a glass bottom dish with saline, that is set above either a 20X or 60X water immersion

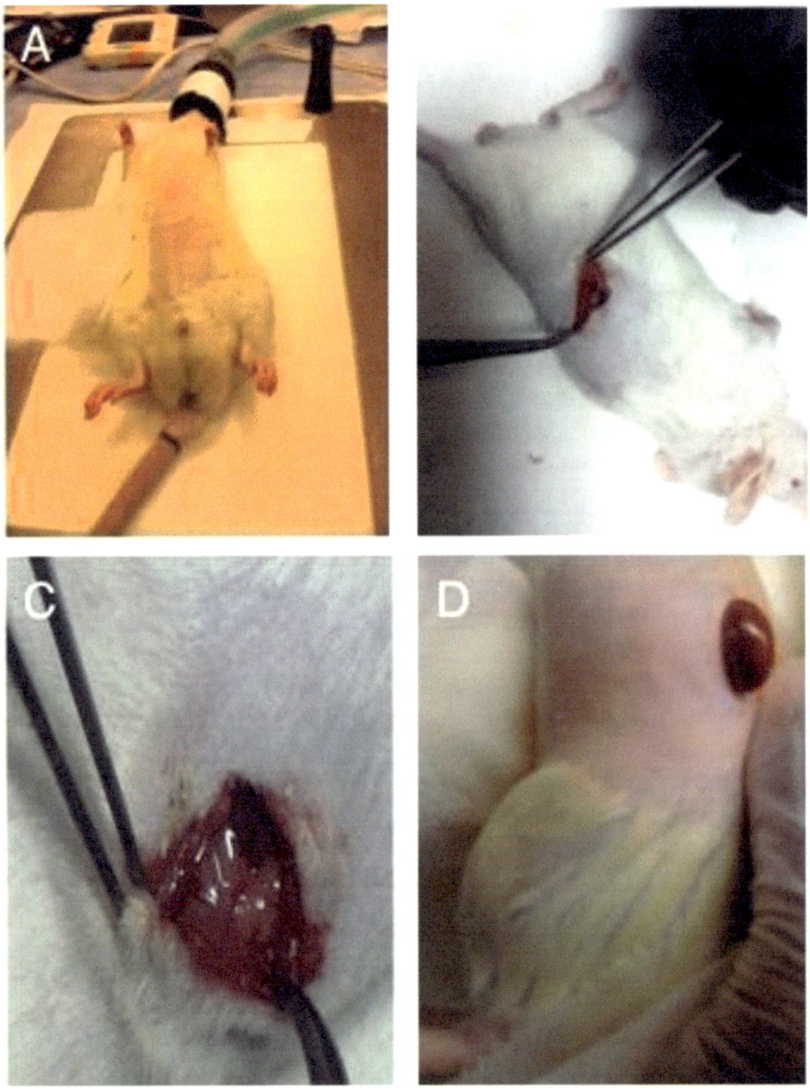

Figure 2.
Digital images illustrate common surgical procedures applied to exteriorize the rat kidney for intravital imaging. These images transition from the provision of inhaled anesthesia that sedates the animal and prepares it for surgery (A); to the generation of a flank incision that facilitates the exteriorization of the left kidney (B through D). Digital zoom provides greater detail in image (C) of the field presented in image (B) by accentuating the tuft of perirenal fat situated at the apex of the kidney that is used to gently birth the kidney from the flank incision, as presented in image (D).

objective for imaging, with the animal's body acting as a weight to stabilize kidney position [16, 65, 66] (**Figure 3**).

Fluorescent images can be acquired from externalized organs. Then, measurements can follow an Olympus FV 1000- MPE Microscope set with a Spectra-Physics Mai Tai Deep See laser, tunable from 710 to 990 nm, with dispersion compensation for two-photon microscopy (Olympus Corporation, Tokyo, Japan). The system in question is generally accompanied by a pair of external detectors for multiphoton imaging and dichroic mirrors for collecting blue, green, and red emissions. The emitted light is

DOI: http://dx.doi.org/10.5772/intechopen.107385

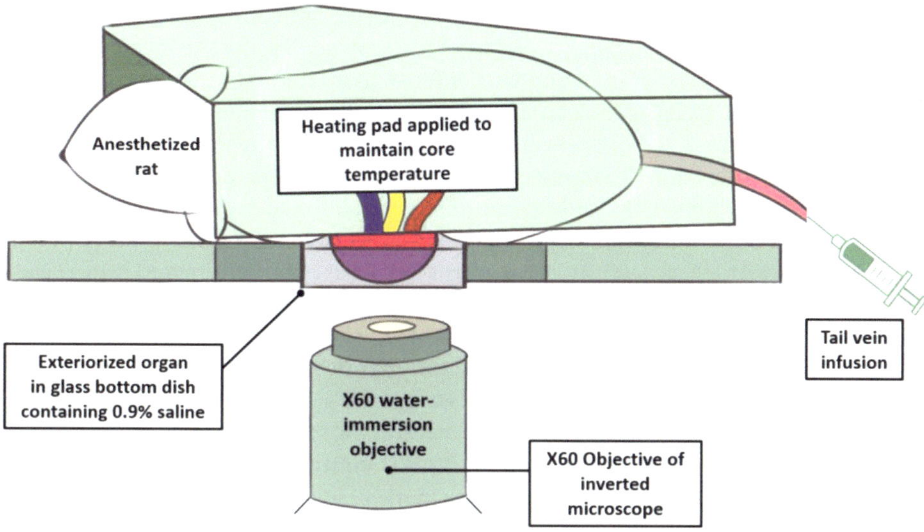

Figure 3.
A schematic illustrating the renal IVM process. In this image, an anesthetized rat, covered with a heating pad to maintain core temperature, has its left kidney exteriorized and placed in a 50 mm glass-bottom dish, filled with saline, and set above the stage of an inverted microscope with a Nikon ×60 1.2-NA water-immersion objective. A 25-gauge butterfly needle was inserted into the dilated tail vein and attached to a syringe containing injectates.

collected using a 3-fixed bandpass filter system: 420–460 nm (blue channel), 495–540 nm (green channel), and 575–630 nm (red channel). The system can be mounted on an Olympus IX81 inverted microscope for conducting live imaging.

5. Renal gene and cell therapies and evolving tissue and engineering techniques devised using fluorescently-tagged dextrans

Concurrently, advances in renal IVM have brought several advantages over in vitro and ex vivo models by providing real-time assessments of dynamic processes at the cellular and subcellular levels. These advances have relied on dextrans to optimize exogenous gene/cell delivery methods, tissue functionality following these alterations, and whole bioengineered organ scaffold integrity. Such advances have been applied to support regenerative medicine strategies, namely renal gene and cell therapies, as well as tissue engineering.

5.1 Optimization of exogenous gene/cell delivery methods

Previous studies that were used to verify that hydrodynamic delivery facilitates robust exogenous transgene and cell distribution throughout the live rodent kidney were based on the internalization of low-, intermediate-, and high-molecular-weight exogenous macromolecules. These infusates were comparable to transgene and transcell suspension [17]. Such studies presented overwhelming evidence that hydrodynamic injections augmented with vascular cross clamps can consistently induce cellular uptake of exogenous low, intermediate, and large macromolecules in numerous live nephron segments.

Interestingly, this rapid injection method supported the robust apical cellular internalization of large-molecular-weight dextran molecules like the incorporation of low-molecular-weight dextran, along with intense basolateral distributions. Data also provided evidence that large molecular-weight dextran molecules could atypically access the tubule lumen at high concentrations after being delivered to the kidney. Likewise, the introduction of large molecular weight, 150-kDa, dextran molecules into the vasculature before hydrodynamic delivery facilitated their internalization within tubular epithelial cells after the injection process was conducted using isotonic saline. Nevertheless, such atypical access to tubular epithelium and lumen was transient and appeared to only occur as a consequence of the hydroporation process. Such results highlight possible delivery routes of transgene entry that can support renal genetic transformation induced by hydrodynamic injections.

These versatile and fluorescently-tagged molecules also helped examine correlations between hydrodynamic injection parameters and reliable transgene expression/cellular incorporation. Specifically, the conditions required to infuse the transgenes at various injection rates were examined to provide insight into each infusion rate's effectiveness and impact on renal structure and function. The resulting data were compared to standard systemic fluid delivery to the kidney in normal rats via jugular and tail vein infusions. The high molecular weight (150-kDa) fluorescent dextran molecules were delivered systemically via either venous route and were characteristically restricted within the peritubular capillaries surrounding intact proximal and distal tubules. This probe was widely distributed within the vasculature of nephron segments that were accessible for imaging by our two-photon microscope independent of the infusion method. This imaging technique can extensively survey the distribution of the fluorescent dye as a function of renal tissue depth.

5.2 Examination of tissue functionality following exogenous gene/cell delivery

In comparison, after identifying a time course for renal recovery and viable delivery, it was necessary to investigate whether these processes would hinder intrinsic renal structural and functional capacities. For this, systemically introduced fluorescent low/large molecular weight dextrans helped investigate the potential uptake and distribution of dyes in superficial nephron cross sections in animals that received hydrodynamic retrograde injections.

Studies were presented to confirm whether this gene delivery method can induce significant degrees of injury. For such studies, morphological and functional assays were conducted 3–28 days post-non-viral and viral hydrodynamic injections to examine microvascular integrity and metabolically activity after gene delivery and expression. Rats received jugular/tail vein infusions of 150-kDa and 3-kDa dextrans to detect augmentations to native renal filtration and endocytic uptake capacities. These studies revealed the maintained innate capacities in each case, as the 3-kDa dye was rapidly filtered and endocytosed by the proximal tubule epithelia, and the 150-kDa dye was retained within the peritubular vasculature. Such results are consistent with normal renal function [12, 16, 17]. Nephron structure and function appeared normal after hydrodynamic delivery, transgene expression, and cellular incorporation.

A significant result from these experiments revealed that the low-molecular-weight dextran molecules were taken up equally well by cells that either did or did not express fluorescent proteins in rats treated with various types of transgene vectors (plasmid and adenovirus vectors). This characteristic indicated that these cells retained functional activity. Again, the data are consistent with endocytic uptake of low molecular weight dextran molecules in rat kidneys [9, 12, 13, 16, 17, 65]. These observations showed that nephron segments could retain vital functional capacities after rapid fine-needle, hydrodynamic venous delivery. Alternatively, this cell delivery technique can also be used to establish tumor models in this organ, thereby providing an extension of its utility [68].

5.3 Evaluation of whole bioengineered organ scaffold integrity

Efforts in tissue engineering have heightened the desire for alternatives such as the bioartificial kidney [69–72]. Whole organ decellularization has been described as one of the most promising ways of constructing a bioartificial kidney [73]. Decellularization focuses on extracting the extracellular matrix (ECM) from the native kidney with as many structural and functional clues as possible. The ECM can then be employed as a natural template for regeneration, as observed traditionally in commercial substitutes [74].

This technique has gained much attention lately, yet maintaining adequate scaffold integrity in the post-transplantation environment remains a considerable challenge. Specifically, there is still a limited understanding of scaffold responses post-transplantation and ways we can improve scaffold durability to withstand the in vivo environment. Recent studies have outlined vascular events that limit organ scaffold viability for long-term transplantation. However, these insights have relied on in vitro/in vivo approaches that lack adequate spatial and temporal resolutions to investigate such issues at the microvascular level.

As a result, intravital microscopy has been recently used to gain instant feedback on their structure, function, and deformation dynamics [75]. This process was able to capture the effects of in vivo blood flow on the decellularized glomerulus, peritubular capillaries, and tubules after autologous orthotopic transplantation into rats. Large molecular weight dextran molecules labeled the vasculature. They revealed substantial degrees of translocation from glomerular and peritubular capillary tracks to the decellularized tubular epithelium and lumen as early as 12 hours after transplantation, providing real-time evidence of the increases in microvascular permeability. Macromolecular extravasation persisted for a week, during which the decellularized microarchitecture was significantly compromised and thrombosed. These results indicate that in vivo multiphoton microscopy is a powerful approach for studying scaffold viability and identifying ways to promote scaffold longevity and vasculogenesis in bioartificial organs.

6. Conclusions

Dextrans are widely used molecules for various applications within medicine. Traditionally, these complex branched glucans have been used as plasma expanding and antithrombotic agents. In more recent times, their applications have extended to the realm of regenerative medicine, where researchers have found niche roles in developing and evaluating the delivery of novel cell- and gene- therapeutics and imaging agents for in vivo investigations. Developments in optical microscopy have also aligned with these applications to produce exciting opportunities for renal intravital models that have now extended to the field of whole organ bioengineering. As a result, the chapter was used to provide insights into how optical microscopic techniques have advanced kidney imaging, the use of fluorescent dextrans, and intravital microscopy. This approach was supported by illustrating how a well-established delivery technique supports exogenous gene/cell delivery to the kidney, along with an example of whole organ bioengineering techniques that can be evaluated using IVM. The illustration can be accessed in further detail in the literature. Overall, it is hoped that this chapter will support future regenerative and bioengineering efforts by emphasizing the relevant methodologies needed to conduct these intravital studies.

Acknowledgements

The author acknowledges funding from the College of Medicine at Khalifa University, and Grant Numbers: FSU-2020-2025 and RC2-2018-022 (HEIC), as well as Ms. Raheema Khan for her help compiling articles to present in this chapter.

Conflict of interest

The author declares no conflict of interest.

Author details

Peter R. Corridon[1,2,3]

1 Biomedical Engineering and Healthcare Engineering Innovation Center, Khalifa University, Abu Dhabi, UAE

2 Department of Immunology and Physiology, College of Medicine and Health Sciences, Khalifa University, Abu Dhabi, UAE

3 Center for Biotechnology, Khalifa University, Abu Dhabi, UAE

*Address all correspondence to: peter.corridon@ku.ac.ae

References

[1] Neu B, Wenby R, Meiselman HJ. Effects of dextran molecular weight on red blood cell aggregation. Biophysical Journal. 2008;**95**(6):3059-3065

[2] Isbister JP, Fisher MM. Adverse effects of plasma volume expanders. Anaesthesia and Intensive Care. 1980;**8**(2):145-151

[3] Mays T, Mays T. Intravenous iron-dextran therapy in the treatment of anemia occurring in surgical, gynecologic and obstetric patients. Surgery, Gynecology & Obstetrics. 1976;**143**(3):381-384

[4] Huskisson L. Intravenous volume replacement: Which fluid and why? Archives of Disease in Childhood. 1992;**67**(5):649-653

[5] Tinawi M. New trends in the utilization of intravenous fluids. Cureus. 2021;**13**(4):e14619

[6] Ukkund SJ et al. 1 - dextran nanoparticles: Preparation and applications. In: Venkatesan J et al., editors. Polysaccharide Nanoparticles. Amsterdam, Netherlands: Elsevier; 2022. pp. 1-31

[7] Terg R et al. Pharmacokinetics of Dextran-70 in patients with cirrhosis and ascites undergoing therapeutic paracentesis. Journal of Hepatology. 1996;**25**(3):329-333

[8] Sandoval RM, Molitoris BA. Quantifying glomerular permeability of fluorescent macromolecules using 2-photon microscopy in Munich Wistar rats. Journal of Visualized Experiments. 2013;**74**:e50052:1-7

[9] Dunn KW, Sandoval RM, Molitoris BA. Intravital imaging of the kidney using multiparameter multiphoton microscopy. Nephron. Experimental Nephrology. 2003;**94**(1):e7-e11

[10] Hato T, Winfree S, Dagher PC. Intravital imaging of the kidney. Methods. 2017;**128**:33-39

[11] Hato T, Winfree S, Dagher PC. Kidney Imaging: Intravital Microscopy. Methods in Molecular Biology. 2018;**1763**:129-136

[12] Ashworth SL et al. Two-photon microscopy: Visualization of kidney dynamics. Kidney International. 2007;**72**(4):416-421

[13] Dunn KW et al. Functional studies of the kidney of living animals using multicolor two-photon microscopy. American Journal of Physiology. Cell Physiology. 2002;**283**(3):C905-C916

[14] Sandoval RM, Molitoris BA. Intravital multiphoton microscopy as a tool for studying renal physiology and pathophysiology. Methods. 2017;**128**:20-32

[15] Tanner GA. Glomerular sieving coefficient of serum albumin in the rat: A two-photon microscopy study. American Journal of Physiology. Renal Physiology. 2009;**296**(6):F1258-F1265

[16] Corridon PR et al. Intravital imaging of real-time endogenous actin dysregulation in proximal and distal tubules at the onset of severe ischemia-reperfusion injury. Scientific Reports. 2021;**11**(1):8280

[17] Corridon PR et al. A method to facilitate and monitor expression of exogenous genes in the rat kidney using plasmid and viral vectors. American Journal of Physiology. Renal Physiology. 2013;**304**(9):F1217-F1229

[18] Mehvar R. Dextrans for targeted and sustained delivery of therapeutic and imaging agents. Journal of Controlled Release. 2000;**69**(1):1-25

[19] Vaghela R et al. Actually seeing what is going on – Intravital microscopy in tissue engineering. Frontiers in Bioengineering and Biotechnology. 2021;**9**. DOI: 10.3389/fbioe.2021.627462

[20] Jun YW et al. Addressing the autofluorescence issue in deep tissue imaging by two-photon microscopy: The significance of far-red emitting dyes. Chemical Science. 2017;**8**(11):7696-7704

[21] Small DM et al. Multiphoton fluorescence microscopy of the live kidney in health and disease. Journal of Biomedical Optics. 2014;**19**(2):020901

[22] Hall AM et al. In vivo multiphoton imaging of mitochondrial structure and function during acute kidney injury. Kidney International. 2013;**83**(1):72-83

[23] Schuh CD et al. Long wavelength multiphoton excitation is advantageous for intravital kidney imaging. Kidney International. 2016;**89**(3):712-719

[24] Masters BR. Correlation of histology and linear and nonlinear microscopy of the living human cornea. Journal of Biophotonics. 2009;**2**(3):127-139

[25] Schmitt JM, Knuttel A, Yadlowsky M. Confocal microscopy in turbid media. Journal of the Optical Society of America. A, Optics, Image Science, and Vision. 1994;**11**(8):2226-2235

[26] Helmchen F, Denk W. Deep tissue two-photon microscopy. Nature Methods. 2005;**2**(12):932-940

[27] Denk W, Svoboda K. Photon upmanship: Why multiphoton imaging is more than a gimmick. Neuron. 1997;**18**(3):351-357

[28] Dunn KW, Young PA. Principles of multiphoton microscopy. Nephron. Experimental Nephrology. 2006;**103**(2):e33-e40

[29] Comper WD, Haraldsson B, Deen WM. Resolved: Normal glomeruli filter nephrotic levels of albumin. Journal of the American Society of Nephrology. 2008;**19**(3):427-432

[30] Kang JJ et al. Quantitative imaging of basic functions in renal (patho)physiology. American Journal of Physiology. Renal Physiology. 2006;**291**(2):F495-F502

[31] Wang E et al. Rapid diagnosis and quantification of acute kidney injury using fluorescent ratio-metric determination of glomerular filtration rate in the rat. American Journal of Physiology. Renal Physiology. 2010;**299**(5):F1048-F1055

[32] Medarova Z et al. In vivo imaging of siRNA delivery and silencing in tumors. Nature Medicine. 2007;**13**(3):372-377

[33] Vaghela R et al. Actually seeing what is going on - Intravital microscopy in tissue engineering. Frontiers in Bioengineering and Biotechnology. 2021;**9**:627462

[34] Weigert R et al. Intravital microscopy: A novel tool to study cell biology in living animals. Histochemistry and Cell Biology. 2010;**133**(5):481-491

[35] Molitoris BA, Sandoval RM. Intravital multiphoton microscopy of dynamic renal processes. American Journal of Physiology. Renal Physiology. 2005;**288**(6):F1084-F1089

[36] Giampetraglia M, Weigelin B. Recent advances in intravital microscopy for preclinical research. Current Opinion in Chemical Biology. 2021;**63**:200-208

[37] Mihlan M et al. Surprises from Intravital imaging of the innate immune response. Annual Review of Cell and Developmental Biology. 2022;**38**. DOI: 10.1146/annurev-cellbio-120420-112849

[38] Taqueti VR, Jaffer FA. High-resolution molecular imaging via intravital microscopy: Illuminating vascular biology in vivo. Integrative Biology. 2013;**5**(2):278-290

[39] Shaya J et al. Design, photophysical properties, and applications of fluorene-based fluorophores in two-photon fluorescence bioimaging: A review. Journal of Photochemistry and Photobiology C: Photochemistry Reviews. 2022;**52**:100529

[40] Cai N et al. Recent advances in fluorescence recovery after Photobleaching for decoupling transport and kinetics of biomacromolecules in cellular physiology. Polymers (Basel). 2022;**14**(9). DOI: 10.3390/polym14091913

[41] Dhawan AP, D'Alessandro B, Fu X. Optical imaging modalities for biomedical applications. IEEE Reviews in Biomedical Engineering. 2010;**3**:69-92

[42] Pachekrepapol U, Horne DS, Lucey JA. Effect of dextran and dextran sulfate on the structural and rheological properties of model acid milk gels. Journal of Dairy Science. 2015;**98**(5):2843-2852

[43] Lencer WI et al. FITC-dextran as a probe for endosome function and localization in kidney. The American Journal of Physiology. 1990;**258**(2 Pt 1): C309-C317

[44] Ruben M, Sandoval & Bruce A, Molitoris BA. Fluorescent Dextrans in Intravital multi-photon microscopy. In: Advances in Intravital Microscopy: From Basic to Clinical Research. Weigert R. Editor. Dordrecht: Springer Netherlands; 2014. pp. 205-219.

[45] Lamke LO, Liljedahl SO. Plasma volume changes after infusion of various plasma expanders. Resuscitation. 1976;**5**(2):93-102

[46] Miao KH, Guthmiller KB. Dextran. Treasure Island (FL): StatPearls; 2022

[47] Ghiran IC. Introduction to fluorescence microscopy. Methods in Molecular Biology. 2011;**689**:93-136

[48] Valeur B, Berberan-Santos MN. A brief history of fluorescence and phosphorescence before the emergence of quantum theory. Journal of Chemical Education. 2011;**88**(6):731-738

[49] Marsh PN, Burns D, Girkin JM. Practical implementation of adaptive optics in multiphoton microscopy. Optics Express. 2003;**11**(10):1123-1130

[50] Lakowicz JR et al. Time-resolved fluorescence spectroscopy and imaging of DNA labeled with DAPI and Hoechst 33342 using three-photon excitation. Biophysical Journal. 1997;**72**(2 Pt 1): 567-578

[51] de Grauw CJ et al. Imaging properties in two-photon excitation microscopy and effects of refractive-index mismatch in thick specimens. Applied Optics. 1999;**38**(28):5995-6003

[52] Rothman SS. Lessons from the Living Cell : The Culture of Science and the Limits of Reductionism. New York: McGraw-Hill; 2002

[53] Stutzmann G. Seeing the brain in action: How multiphoton imaging has advanced our understanding of neuronal function. Microscopy and Microanalysis. 2008;**14**(6):482-491

[54] Chia TH et al. Multiphoton fluorescence lifetime imaging of intrinsic fluorescence in human and rat brain tissue reveals spatially distinct NADH binding. Optics Express. 2008;**16**(6):4237-4249

[55] Bacskai BJ et al. Four-dimensional multiphoton imaging of brain entry, amyloid binding, and clearance of an amyloid-beta ligand in transgenic mice. Proceedings of the National Academy of Sciences of the United States of America. 2003;**100**(21):12462-12467

[56] Lin J et al. Assessment of liver steatosis and fibrosis in rats using integrated coherent anti-stokes Raman scattering and multiphoton imaging technique. Journal of Biomedical Optics. 2011;**16**(11):116024

[57] Thorling CA et al. Multiphoton microscopy and fluorescence lifetime imaging provide a novel method in studying drug distribution and metabolism in the rat liver in vivo. Journal of Biomedical Optics. 2011;**16**(8):086013

[58] Peti-Peterdi J, Burford JL, Hackl MJ. The first decade of using multiphoton microscopy for high-power kidney imaging. American Journal of Physiology. Renal Physiology. 2012;**302**(2):F227-F233

[59] Hall AM et al. Multiphoton imaging of the functioning kidney. Journal of the American Society of Nephrology. 2011;**22**(7):1297-1304

[60] Molitoris BA, Sandoval RM. Multiphoton imaging techniques in acute kidney injury. Contributions to Nephrology. 2010;**165**:46-53

[61] Liang X, Graf BW, Boppart SA. In vivo multiphoton microscopy for investigating biomechanical properties of human skin. Cellular and Molecular Bioengineering. 2011;**4**(2):231-238

[62] Masters B, So P. Confocal microscopy and multi-photon excitation microscopy of human skin in vivo. Optics Express. 2001;**8**(1):2-10

[63] Masters BR, So PT. Multi-photon excitation microscopy and confocal microscopy imaging of In vivo human skin: A comparison. Microscopy and Microanalysis. 1999;**5**(4):282-289

[64] Molitoris BA, Sandoval RM. Techniques to study nephron function: Microscopy and imaging. Pflügers Archiv. 2009;**458**(1):203-209

[65] Dunn KW, Sutton TA, Sandoval RM. Live-animal imaging of renal function by multiphoton microscopy. Current Protocols in Cytometry. 2007. Chapter 12

[66] Tanner GA et al. Micropuncture gene delivery and intravital two-photon visualization of protein expression in rat kidney. American Journal of Physiology. Renal Physiology. 2005;**289**(3):F638-F643

[67] Kolb AL et al. Exogenous gene transmission of Isocitrate dehydrogenase 2 mimics ischemic preconditioning protection. Journal of the American Society of Nephrology. 2018;**29**(4):1154-1164

[68] Li J, Yao Q, Liu D. Hydrodynamic cell delivery for simultaneous establishment of tumor growth in mouse lung, liver and kidney. Cancer Biology & Therapy. 2011;**12**(8):737-741

[69] Corridon PR et al. Bioartificial Kidneys. Current Stem Cell Reports. 2017;**3**(2):68-76

[70] Kim S et al. Current strategies and challenges in engineering a bioartificial kidney. Frontiers in Bioscience (Elite Edition). 2015;7:215-228

[71] Corridon PR. In vitro investigation of the impact of pulsatile blood flow on the vascular architecture of decellularized porcine kidneys. Sci Rep. 2021;**11**:16965. DOI: 10.1038/s41598-021-95924-5

[72] Pantic IV, et al. Analysis of vascular architecture and parenchymal damage generated by reduced blood perfusion in decellularized porcine kidneys using a gray level co-occurrence matrix. Frontiers in Cardiovascular Medicine, 2022;**9**:797283

[73] Gilpin A, Yang Y. Decellularization strategies for regenerative medicine: From processing techniques to applications. BioMed Research International. 2017;**2017**:9831534

[74] Corridon P et al. Time-domain terahertz spectroscopy of artificial skin. SPIE BiOS. SPIE. 2006:**6080**

[75] Corridon PR. Intravital microscopy datasets examining key nephron segments of transplanted decellularized kidneys. Scientific Data. 2022;**9**(1):561. DOI: 10.1038/s41597-022-01685-9

Chapter 2

Application of Excitation-Emission Matrix Fluorescence (EEMF) in the Wastewater Field

Francisco Rodríguez-Vidal

Abstract

Fluorescence is a versatile and useful analytical technique for the analysis of waters, both natural waters (freshwaters and marine waters) and wastewaters (urban wastewaters and industrial effluents). Among the various fluorescence techniques currently available, excitation-emission matrix fluorescence (EEMF) is the most used nowadays since it provides comprehensive information on the dissolved organic matter (DOM) present in water. EEMF spectra can be represented either in the form of a 3D-graph or a 2D-contour map and fluorescence peaks can be studied by the fast and simple peak-picking method (more suitable for routine measurements in water treatment plants, allowing a rapid response in case of potential problems in the sequence of treatment) or using mathematical tools such as PARAFAC (more suitable for research purposes and accurate identification of the fluorophores). The EEMF peaks commonly found in waters are peaks A and C (humic substances), peaks B1, B2, T1, and T2 (protein-like peaks), and peak M (microbial-like peak). EEMF was first applied to the characterization of natural waters, but in recent years, more attention is being paid to the wastewater field. Urban wastewaters have been mostly studied, whereas there are fewer studies focused on industrial effluents. This chapter provides a brief review of these EEFM applications.

Keywords: excitation-emission matrix fluorescence (EEMF), natural waters, wastewaters, industrial effluents, humic substances

1. Introduction

In recent years, the number of studies using fluorescence techniques for the characterization of dissolved organic matter (DOM) in natural and wastewaters has significantly increased [1–16]. There are several reasons for this fact: fluorescence is a fast, sensitive, and nondestructive analytical technique that requires small volumes of the sample. Moreover, in most cases, samples just require a simple pretreatment (pH adjustment and filtration, if necessary) and fluorescence probes can be readily adapted to automated devices for in situ measurements.

Fluorescence offers several advantages over other alternative techniques often used in water analysis. For instance, global parameters such as biochemical oxygen demand (BOD) and chemical oxygen demand (COD) provide no information on the structure and properties of DOM, in addition to being time-consuming methods (5 days and 2 hours, respectively). Other more sophisticated techniques, such as gas chromatography-mass spectrometry (GC/MS), infrared spectroscopy (FTIR), and ^{1}H- and ^{13}C-nuclear magnetic resonance (NMR), require complicated and laborious procedures for extraction-purification of the aqueous samples. Moreover, when analyzing complex matrices, such as wastewaters, FTIR, and NMR signals, usually overlap into broad and poorly resolved bands, thus making the interpretation of the spectra difficult [17].

Several fluorescence techniques can be applied to the analysis of freshwaters and wastewaters, such as the conventional emission scan fluorescence (ESF) and the more interesting synchronous fluorescence spectroscopy (SFS). However, the most useful and complete technique used at present is excitation-emission matrix fluorescence (EEMF, also known as total luminescence spectroscopy: TLS), in which a series of emission scans are collected for a range of excitation wavelengths. The generated matrix of data can be represented either in the form of a 3D-graph or a 2D-contour map, thus making it easier for quick identification of the main fluorescence peaks present in the sample. Peak coordinates are represented as ($\lambda_{ex}/\lambda_{em}$, in nm) and their maximum intensities of fluorescence (F_{max}) are representative of the relative concentration of the fluorophores [18, 19]. The interpretation of the spectra can be conducted using either the traditional "peak-picking" method or more sophisticated mathematical tools, such as the parallel factor analysis (PARAFAC). The peak-picking method is simpler, faster (an EEMF spectrum in the range $\lambda_{ex}/\lambda_{em}$: 220–450/300–550 nm is usually collected in 8–10 minutes), and useful for quick monitoring of the typical fluorophores present in waters, which makes it more suitable for routine measurements in water treatment plants as these plants usually demand fast and user-friendly analytical techniques. PARAFAC turns out to be more appropriate for research purposes since this tool requires a higher degree of analytical expertise.

The main EEMF peaks found in natural and wastewaters are the following and can be classified into three major groups ($\lambda_{ex}/\lambda_{em}$, in nm):

- Humic-like peaks: peak A (230–260/400–480 nm, fulvic-like), peak C (320–360/420–460 nm, humic-like). They are associated with humic substances (fulvic and humic acids).

- Protein-like peaks: tyrosine-like peaks B1 (275–310/305–320 nm) and B2 (220–237/305–320 nm), tryptophan-like peaks T1 (275–285/320–350 nm), and T2 (215–237/340–381 nm). They are mainly associated with the presence of proteinaceous material (proteins and peptides) containing the amino acids tyrosine and tryptophan. However, this fluorescence might not be due exclusively to proteins, since recent studies have reported that humic substances can encapsulate proteins under certain circumstances, indicating a potential combination between them. Additionally, some polyphenolic compounds, such as lignin, have also been reported to exhibit tryptophan-like fluorescence [20].

- Microbial-like peak M (290–310/370–420 nm): this peak is associated with the release of organic compounds from recent microbiological activity.

Unfortunately, neither lipids (oil and grease) nor carbohydrates (both of them usually present in wastewaters) can be detected by EEMF, which constitutes a drawback when a comprehensive characterization of the water is required.

A location of these peaks in a typical EEMF spectrum is shown in **Figure 1**. In addition to the aforementioned peaks, several fluorescence indices are also used in some studies for specific purposes (see **Figure 1**), such as:

Fluorescence index (FI), first introduced by McKnight [21], is calculated as the ratio of emission intensity at 450/500 nm measured at λ_{ex} = 370 nm:

$$FI = I_{Em\ 450}/I_{Em\ 500}, \text{at } \lambda_{ex} = 370\ \text{nm} \quad (1)$$

This index has been mostly used to elucidate the origin of fulvic acids in freshwaters (FI values around 1.9 denote fulvic acids of microbial origin, whereas values around 1.4 indicate terrestrially derived fulvic acids [22]. FI has been also reported to show a negative correlation with the aromaticity of humic substances [23].

Humification index (HIX): this index was proposed by Zsolnay [24] and is determined as the ratio of fluorescence intensities of the integrated emission region of λ_{em} = 435–480 nm divided by that of λ_{em} = 300–345 nm, measured at λ_{ex} = 254 nm.

$$HIX = \sum I_{Em\ 435\text{–}480}/\sum I_{Em\ 300\text{–}345}, \text{at } \lambda_{ex} = 254\ \text{nm} \quad (2)$$

Later on, a modification of the original HIX was introduced, calculated as the emission intensity in the 435–480 nm region divided by the sum of total intensities in the (300–345 + 435–480) nm regions. This index is denoted as "normalized HIX" (HIX_{norm}), as it ranges from 0 to 1.

$$HIX_{norm} = \sum I_{Em\ 435\text{–}480}/\left(\sum I_{Em\ 300\text{–}345} + \sum I_{Em\ 435\text{–}480}\right), \text{at } \lambda_{ex} = 254\ \text{nm} \quad (3)$$

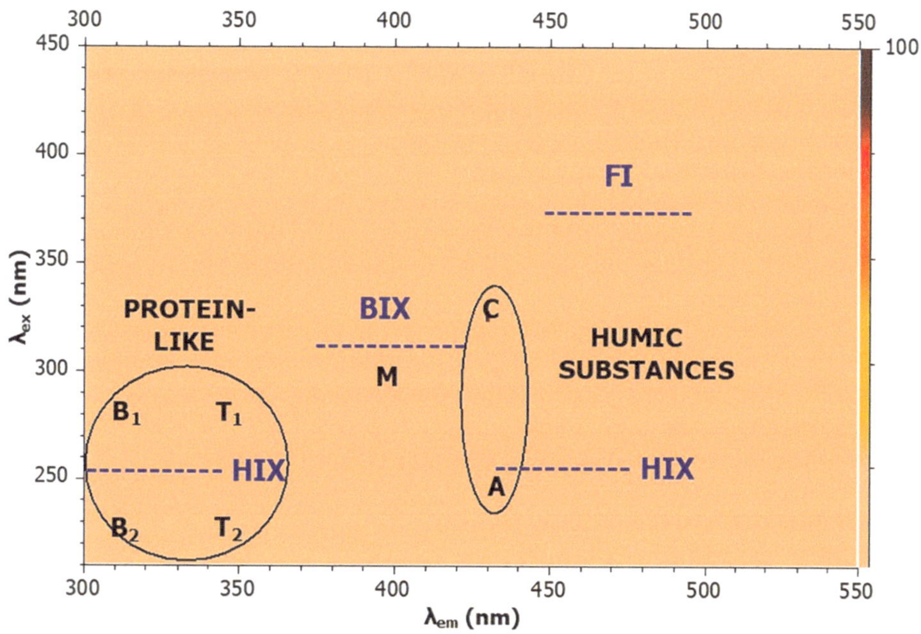

Figure 1.
Location of the main EEMF peaks and fluorescence indices in waters.

HIX is related to the degree of humification of the organic matter in waters and is strongly correlated with DOM (dissolved organic matter) aromaticity [25].

Biological index (BIX), first introduced by Huguet [26], is determined by dividing the fluorescence intensities at the emission wavelengths of 380 and 430 nm, measured at λ_{ex} = 310 nm:

$$BIX = I_{Em\ 380}/I_{Em\ 430}, \text{at } \lambda_{ex} = 310 \text{ nm} \tag{4}$$

As shown in **Figure 1**, BIX is strongly correlated with peak M, indicating the presence of organic matter recently released by microorganisms in water (autochthonous DOM from biological origin) [23].

2. Application of fluorescence in water analysis

2.1 Fluorescence and natural waters

Before getting into the fluorescence applications in the wastewater field, it is interesting to do a brief review of its applications in natural waters, both freshwaters (rivers, reservoirs, etc.) and marine waters since this field has been the most studied for many years. The most abundant EEMF peaks found in natural waters are humic-like peaks (both A and C), which is indicative of the presence of humic and fulvic acids in water, the latter constituting the majority fraction of the aquatic humic substances. Actually, a considerable presence of protein-like peaks in freshwaters is usually related to wastewater discharges of anthropogenic origin [18, 27]. Humic substances make up most of the NOM (around 30–50%) present in freshwaters [28] and are originated from both humification processes occurring during the decomposition of vegetable organic matter in water (autochthonous microbial origin) and elutriation of soil humic substances from the surrounding terrain (terrestrial origin).

There are several drawbacks directly related to an excessive presence of humic substances in water, such as an increased formation of disinfection by-products upon chlorination (mainly trihalomethanes), they can act as carriers for micropollutants and heavy metal ions via the formation of soluble complexes with them, they contribute to membrane fouling in membrane-based water treatments (for instance, membrane biological reactors or MBR), they contribute to the biofilm formation in water distribution pipelines and they can hinder the adsorption of micropollutants onto activated carbon.

EEMF can provide interesting information on humic substances structure and properties: the location and shift of the peaks and their fluorescence intensities are correlated to some parameters, such as the aromaticity degree, carboxylic acidity, and the degree of humification. Additionally, there are several well-established behaviors concerning the fluorescence of humic substances [29, 30], namely:

- the intensity of the fluorescence peaks (both A and C) decreases with increasing the macromolecule molecular size.

- concerning substituted aromatic moieties in the humic macromolecule: electron-donating groups (-OH, $-NH_2$, and $-OCH_3$) cause an increase in the fluorescence intensity, whereas electron-withdrawing groups (-COOH) cause the opposite effect.

- hydroxyl, alkoxyl, amino, and carbonyl-containing substituents usually cause a red-shift (fluorescence maxima shift toward longer wavelengths).

- a reduction in the aromaticity degree of the macromolecule (for instance, a reduction in the number of aromatic rings) usually causes a blue-shift (fluorescence maxima shift toward shorter wavelengths).

Figure 2 shows the EEMF spectrum (2D-contour map) of natural water (Úzquiza Reservoir, which supplies to the city of Burgos, Spain) and the EEMF spectrum (3D-graph) of a pure fulvic acid (Nordic fulvic acid, reference material from the international humic substances society). As shown in **Figure 2**, the reservoir water is characterized by only presence of humic-like peaks, a high-intensity peak A (fulvic-like), and a less intense peak C (humic-like). There is no presence of protein-like peaks, which is indicative of the absence of urban wastewater discharges and therefore, a clear sign of good quality water. Obviously, the 3D spectrum of the pure fulvic acid (**Figure 2**) only contains humic-like peaks, being the fulvic-like peak A the majority one.

2.2 Fluorescence and urban wastewaters

Dissolved organic matter (DOM) in wastewater comprises a great variety of organic compounds, from low-molecular weight (MW) substances (amino acids, small organic acids, simple sugars, etc.) to high-MW compounds (proteins, humic substances, carbohydrates, etc.) [23, 31–34]. In the wastewater field, fluorescence has been mostly applied to the characterization of effluent organic matter (EfOM) from urban wastewater treatment plants (WWTPs) [1–4, 7, 8, 35, 36].

Protein-like peaks T1 and T2 (tryptophan-like peaks) are usually the most abundant EEMF peaks found in urban wastewaters. These peaks originated from both proteinaceous material present in the influent (anthropogenic origin) and protein-like compounds released by microorganisms (soluble microbial products: SMP) during the biological treatment stage in WWTPs [19, 37]. Conversely, the presence of tyrosine-like peaks (B1 and B2) in urban wastewaters is less frequent because tyrosine

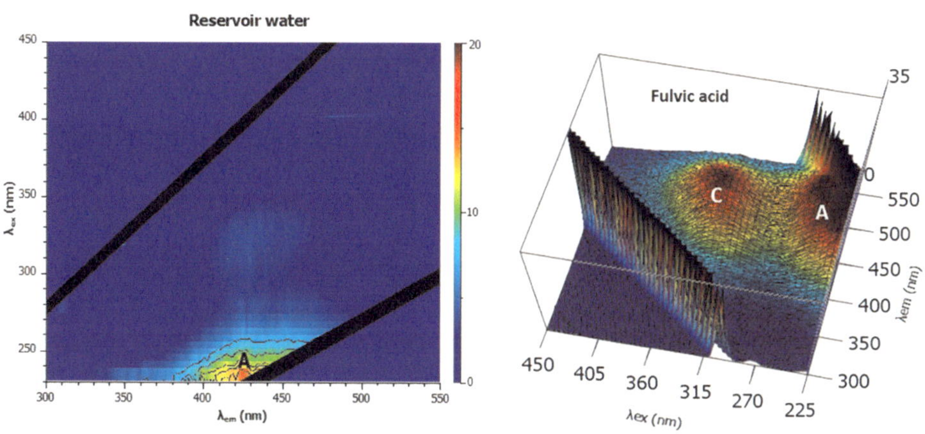

Figure 2.
EEMF spectrum (2D contour plot) of a reservoir water (left) and EEMF spectrum (3D graph) of aquatic fulvic acid (right). The 3D fulvic acid spectrum also shows the first and second order Rayleigh scattering peaks.

fluorescence is usually quenched within high molecular weight proteins due to resonance energy transfer [1]. That is why the detection of peaks B in the EEMF spectrum is usually associated with the presence of free tyrosine or tyrosine-containing small peptides (in which tryptophan is not present) in the sample [38].

The relative abundance of tryptophan-like peaks T1 and T2 (T1/T2 ratio) in the influent depends on the specific type of domestic wastewater and the influence of industrial discharges into the municipal WWTP. Consequently, peak T1 is reported as the most abundant in some studies from the literature [19, 34, 35] whereas peak T2 in others [1, 20, 39].

EEMF has also been proved to be useful to track changes in NOM throughout the sequence of treatment in WWTPs [40]. Protein-like peaks are more biodegradable than humic-like peaks, whereas the latter are more amenable to be removed by sedimentation. Therefore, in WWTPs protein-like peaks show greater percentages of removal at the biological treatment stage, whereas humic-like peaks at the clarification stage [37].

Figure 3 shows the EEMF spectrum of an urban wastewater influent and effluent (wastewater treatment plant of Burgos and Spain). Quenching effects caused by the presence of metal ions in the wastewater (mainly iron) are negligible due to their low concentration levels, usually found in urban wastewaters. As shown in **Figure 3**, tryptophan-like peak T2 is the most abundant in this wastewater and the comparison of fluorescence intensities between the influent and the effluent allows the estimation of removal percentages for each peak.

2.3 Fluorescence and industrial wastewaters

In the wastewater field, most studies reported in the literature have focused on urban/domestic wastewaters, but little attention has been paid to industrial effluents. In addition to the organic compounds typically present in urban wastewaters (see Section 2.2), industrial wastewaters can contain a great diversity of organic pollutants depending on the specific industry sector (phenols, pharmaceuticals, organic solvents, surfactants coming from tank cleaning processes, etc.). For this reason and contrary to urban wastewaters (where a typical EEMF spectrum with a predominance of protein-like peaks is expected in most cases), no standard EEMF spectrum can be associated

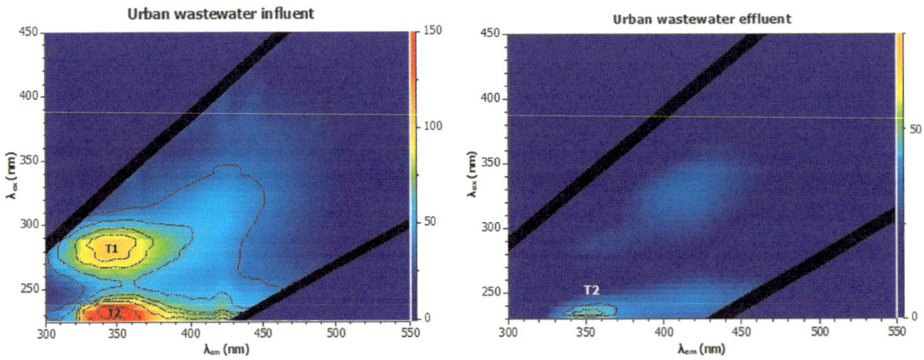

Figure 3.
EEMF spectrum of urban wastewater influent (left) and effluent (right).

with industrial effluents. For instance, food-related industries (milk, brewery, winery, biscuit industries, etc.) do show EEMF spectra similar to those of urban wastewaters (predominance of protein-like peaks) but conversely, old landfill leachates exhibit spectra just containing humic-like peaks: the higher the landfill age (and therefore the higher the humification degree of the humic substances) the greater the humic-like peak C fluorescence intensity [23]. It is interesting to note that some kinds of industries, such as pulp and mill, textile dyeing industries, and slaughterhouses, are reported to potentially show specific fingerprints that could allow a tentative identification of their origin but more research is needed on this issue [23].

Figure 4 shows the EEMF spectrum for a food industry effluent (a cold-meat processing factory) and municipal landfill leachate. As commented earlier, the spectrum of the cold-meat industry effluent is characterized by the predominance of protein-like peaks, whereas that of the landfill leachate exhibits a dominant humic-like fluorescence (peak C), indicating leachate coming from an old landfill.

Table 1 summarizes the different types of water frequently characterized by EEMF along with the references included in this chapter.

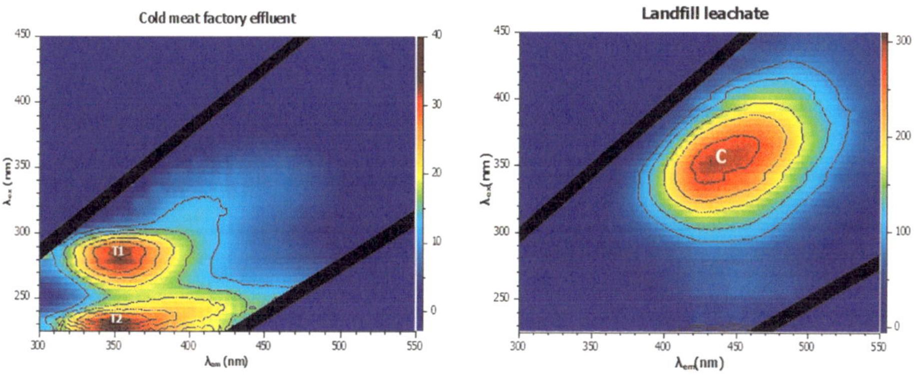

Figure 4.
EEMF spectrum of a food industry wastewater (cold-meat industry effluent) and a municipal landfill leachate.

Type of water		References
Natural waters		[5, 6, 10–12, 14, 16, 21, 24–27, 29, 30, 36, 38]
Urban wastewaters		[1–4, 7–9, 15, 17–20, 23, 28, 31, 33–37, 39]
Industrial wastewaters	Food industries	[32, 41–44]
	Pulp mill industries	[22, 41, 45]
	Textile industries	[41, 46–50]
	Slaughterhouses	[41, 51, 52]
	Landfill leachates	[40, 41, 53–56]
	Pharmaceutical industries	[13]

Table 1.
Types of waters typically analyzed by EEMF and related literature references.

3. Conclusions

Fluorescence, and particularly excitation-emission matrix fluorescence (EEMF), has been proved to be a useful and versatile analytical technique for the characterization of the organic matter present in wastewaters. Due to the fact that fluorescence is a fast and user-friendly technique, it can be easily implemented in wastewater treatment plants for routine measurements, allowing a rapid response to deal with potential problems in the treatment line. New studies in this field are being continuously released and this trend will surely continue in the future.

Author details

Francisco Rodríguez-Vidal
Department of Chemistry, University of Burgos, Burgos, Spain

*Address all correspondence to: qpvito@ubu.es

References

[1] Bridgeman J, Baker A, Carliell C, Carstea E. Determination of changes in wastewater quality through a treatment works using fluorescence spectroscopy. Environmental Technology. 2013;**34**: 3069-3077

[2] Cohen E, Levy GJ, Borisover M. Fluorescent components of organic matter in wastewater: Efficacy and selectivity of the water treatment. Water Research. 2014;**55**:323-334

[3] Guo X, Yu H, Yan Z, Gao H, Zhang Y. Tracking variations of fluorescent dissolved organic matter during wastewater treatment by accumulative fluorescence emission spectroscopy combined with principal component, second derivative and canonical correlation analyses. Chemosphere. 2018;**194**:463-470

[4] Li J, Wang L, Geng L, Xu K, Ren H. Distribution and removal of fluorescent dissolved organic matter in 15 municipal wastewater treatment plants in China. Chemosphere. 2020;**251**:126375

[5] Markechová D, Tomková M, Sádecká J. Fluorescence excitation-emission matrix spectroscopy and parallel factor analysis in drinking water treatment: A review. Polish Journal of Environmental Studies. 2013;**22**:1289-1295

[6] Shutova Y, Baker A, Bridgeman J, Henderson RK. Spectroscopic characterization of dissolved organic matter in drinking water treatment: From PARAFAC analysis to online monitoring wavelengths. Water Research. 2014;**54**:159-169

[7] Xiao K, Yu J, Wang S, Wang Y, Huang X. Relationship between fluorescence excitation-emission matrix properties and the relative degree of DOM hydrophobicity in wastewater treatment effluents. Chemosphere. 2020;**254**:126830

[8] Yu J, Xiao K, Xue W, Wang Y, Huang X. Excitation-emission matrix (EEM) fluorescence spectroscopy for characterization of organic matter in membrane reactors: Principles, methods and applications. Frontiers of Environmental Science & Engineering. 2020;**14**:31

[9] Sciscenko I, Arques A, Micó P, Mora M, García-Ballesteros S. Emerging applications of EEM-PARAFAC for water treatment: A concise review. Chemical Engineering Journal Advance. 2022;**10**:100286

[10] Dai Y, Wang H, Wang J, Wang Z, Ge X. Prediction of water quality based on SVR by fluorescence excitation-emission matrix and UV-Vis absorption spectrum. Spectrochimica Acta – Part A: Molecular and Biomolecular Spectroscopy. 2022; **273**:121059

[11] Fan T, Yao X, Ren H, Zhu H, Zang Y. Multi-spectroscopic investigation of the molecular weight distribution and copper binding ability of dissolved organic matter in Dongping Lake, China. Environmental Pollution. 2022;**300**: 118931

[12] Song Q, Graham N, Tang Y, Kimura K, Yu W. The role of medium molecular weight organics on reducing disinfection by-products and fouling prevention in nanofiltration. Water Research. 2022; **215**:118263

[13] Gharibian S, Hazrati H. Towards practical integration of MBR with electrochemical AOP: Improved biodegradability of real pharmaceutical

wastewater and fouling mitigation. Water Research. 2022;**218**:118478

[14] Zhao Y, Sheng Z, Feng J, Wang X. Relative contributions of different sources in Erhai Lake as revealed by PLS-PM. Chemosphere. 2022;**299**:134377

[15] Xue Z, Lv Z, Liu C, Yu S, Li L. Chromatographic and spectroscopic comparison of dissolved organic matter variation in anaerobic-anoxic-oxic process with tertiary filtration and membrane bioreactor. Journal Water Proceedings Engineering. 2022;**47**: 102693

[16] Jutaporn P, Muenphukhiaw N, Phungsai P, Leungprasert S, Musikavong C. Characterization of DBP precursor removal by magnetic ion exchange resin using spectroscopy and high-resolution mass spectrometry. Water Research. 2022;**217**:118435

[17] Da Silva LA, Flumignan DL, Tininis AG, Pezza HR, Pezza L. Discrimination of brazilian lager beer by H NMR spectroscopy combined with chemometrics. Food Chemistry. 2019; **272**:488-493

[18] Goldman JH, Rounds SA, Needoba JA. Applications of fluorescence spectroscopy for predicting percent wastewater in an urban stream. Environmental Science & Technology. 2012;**46**:4374-4381

[19] Yu H, Song Y, Tu X, Peng J. Assessing removal efficiency of dissolved organic matter in wastewater treatment using fluorescence excitation emission matrices with parallel factor analysis and second derivative synchronous fluorescence. Bioresource Technology. 2013;**144**:595-601

[20] Li WT, Chen SY, Shuang CD, Li AM. Characterization of dissolved organic matter in municipal wastewater using fluorescence PARAFAC analysis and chromatography multi-excitation/ emission scan: A comparative study. Environmental Science & Technology. 2014;**48**:2603-2609

[21] McKnight DM, Byer EW, Westerhoff PK, Doran PT, Kulbe T, Andersen DT. Spectrofluorometric characterization of dissolved organic matter for indication of precursor organic material and aromaticity. Limnology and Oceanography. 2001;**46**: 38-48

[22] Cawley KM, Butler KD, Aiken GR, Larsen LG, Huntington TG, McKnight DM. Identifying fluorescent pulp mill effluent in the Gulf of Maine and its watershed. Marine Pollution Bulletin. 2012;**64**:1678-1687

[23] Rodríguez-Vidal FJ, García-Valverde M, Ortega-Azabache B, González-Martínez A, Bellido-Fernández A. Characterization of urban and industrial wastewaters using excitation-emissión matrix (EEM) fluorescence: Searching for specific fingerprints. Journal of Environmental Management. 2020;**263**: 110396

[24] Zsolnay A, Baigar E, Jimenez M, Steinweg B, Saccomandi F. Differentiating with fluorescence spectroscopy the sources of dissolved organic matter in soils subjected to drying. Chemosphere. 1999;**38**:45-50

[25] Morling K, Herzsprung P, Kamjunke N. Discharge determines production of, decomposition of and quality changes in dissolved organic carbon in pre-dams of drinking water reservoirs. Science Total Environment. 2017;**577**:329-339

[26] Huguet A, Vacher L, Relexans S, Saubusse S, Froidefond JM, Parlanti E. Properties of fluorescent dissolved

organic matter in the Gironde Estuary. Organic Geochemistry. 2009;**40**:706-719

[27] Yu H, Song Y, Du E, Yang N, Peng J, Liu R. Comparison of PARAFAC components of fluorescent disolved and particular organic matter from two urbanized rivers. Environmental Science and Pollution Research. 2016;**23**: 10644-10655

[28] Pernet-Coudrier B, Clouzot L, Varrault G, Mouchel JM. Dissolved organic matter from treated effluent of a major wastewater treatment plant: Characterization and influence on copper toxicity. Chemosphere. 2008;**73**: 593-599

[29] Peuravuori J, Koivikko R, Pihlaja K. Characterization, differentiation and classification of aquatic humic matter separated with different sorbents: Synchronous scanning fluorescence spectroscopy. Water Research. 2002;**36**: 4552-4562

[30] Swietlik J, Sikorska E. Application of fluorescence spectroscopy in the studies of natural organic matter fractions reactivity with chlorine dioxide and ozone. Water Research. 2004;**38**:3791-3799

[31] Esparza-Soto M, Núñez-Hernández S, Fall C. Spectrometric characterization of effluent organic matter of a sequencing batch reactor operated at three sludge retention times. Water Research. 2011;**45**:6555-6563

[32] Janhom T, Wattanachira S, Pavasant P. Characterization of brewery wastewater with spectrofluorometry analysis. Journal of Environmental Management. 2009;**90**:1184-1190

[33] Komatsu K, Onodera T, Kohzu A, Syutsubo K, Imai A. Characterization of dissolved organic matter in wastewater during aerobic, anaerobic and anoxic treatment processes by molecular size and fluorescence analyses. Water Research. 2020;**171**:115459

[34] Yu H, Qu F, Sun L, Liang H, Han Z, Chang H, et al. Relationship between soluble microbial products (SMP) and effluent organic matter (EfOM): Characterized by fluorescence excitation emission matrix coupled with parallel factor analysis. Chemosphere. 2015;**121**: 101-109

[35] Yang L, Shin HS, Hur J. Estimating the concentration and biodegradability of organic matter in 22 wastewater treatment plants using fluorescence excitation emission matrices and parallel factor analysis. Sensors. 2014;**14**: 1771-1786

[36] Yang L, Hur J, Zhuang W. Occurrence and behaviors of fluorescence EEM-PARAFAC components in drinking water and wastewater treatment systems and their applications: A review. Environmental Science and Pollution Research. 2015;**22**: 6500-6510

[37] Rodríguez-Vidal FJ, García-Valverde M, Ortega-Azabache B, González-Martínez A, Bellido-Fernández A, Díez-Blanco V, et al. Monitoring the performance of wastewater treatment plants for organic matter removal using excitation-emission matrix fluorescence. Microchemical Journal. 2022;**175**:107177

[38] Yamashita Y, Tanoue E. Chemical characterization of protein-like fuorophores in DOM in relation to aromatic amino acids. Marine Chemistry. 2003;**82**:255-271

[39] Lee S, Ahn KH. Monitoring of COD as an organic indicator in wastewater and treated effluent by fluorescence excitation-emission (FEEM) matrix

characterization. Water Science and Technology. 2014;**50**:57-63

[40] Wang HW, Li XY, Hao ZP, Sun YJ, Wang YN, Li WH, et al. Transformation of dissolved organic matter in concentrated leachate from nanofiltration during ozone-based oxidation processes (O3, O3/H2O2 and O3/UV). Journal of Environmental Management. 2017;**191**:244-251

[41] Rodríguez F, Ortega B, González A, Bellido A. Comprehensive characterization of industrial wastewaters using EEM fluorescence, FT-IR and ^{1}H NMR techniques. Science Total Environment. 2022;**805**:150417

[42] Cooper J, Antony A, Luiz A, Leslie G. Characterization of dissolved organic matter in fermentation industry effluents and comparison with model compounds. Chemosphere. 2019;**234**: 630-639

[43] Ma YB, Amamcharla JK. Front-face fluorescence spectroscopy combined with chemometrics to detect high proteinaceous matter in milk and whey ultrafiltration permeate. Journal of Dairy Science. 2019;**102**:8756-8767

[44] Bi H, Tang L, Gao X, Jia J, Lv H. Spectroscopic analysis on the binding interaction between tetracycline hydrochloride and bovine proteins beta-casein, alphalactalbumin. Journal of Luminiscence. 2016;**178**:72-83

[45] Ciputra S, Antony A, Phillips R, Richardson D, Leslie G. Comparison of treatment options for removal of recalcitrant dissolved organic matter from paper mil effluent. Chemosphere. 2010;**81**:86-91

[46] Cheng C, Wu J, You L, Khan M. Novel insights into variation of dissolved organic matter during textile wastewater treatment by fluorescence excitation-emission matrix. Chemical Engineering Journal. 2018;**335**:13-21

[47] Cheng C, Liu B, Liu C, Shen J, Wu J. Tracking variation of fluorescent dissolved organic matter during full-scale printing and dyeing wastewater treatment. Chemosphere. 2020;**252**: 126559

[48] Liu B, Wu J, Cheng C, Shen J. Identification of textile wastewater in water bodies by fluorescence excitation-emission matrix – parallel factor analysis and high performance size exclusion chromatography. Chemosphere. 2019; **216**:617-623

[49] Wang WL, Cai YZ, Hu HY, Wu QY. Advanced treatment of bio-treated dyeing and finishing wastewater using ozone-biological activated carbon: A study on the synergistic effects. Chemical Engineering Journal. 2019;**359**: 168-175

[50] Qian FY, Sun XB, Liu YD. Effect of ozone on removal of dissolved organic matter and its biodegradability and adsorbability in biotreated textile effluents. Ozone Science and Engineering. 2013;**35**:7-15

[51] Louvet JN, Homeky B, Casellas M, Pons MN, Dagot C. Monitoring of slaughterhouse wastewater biodegradation in a SBR using fluorescence and UV-Visible absorbance. Chemosphere. 2013;**91**:648-655

[52] Wang Y, Li C, Mo H, Sun Y, Chen Y, Wan P. Extraction of chemical fingerprint in food industry wastewater. Advance Material Research. 2013;**726–731**:1484-1490

[53] Liu Z, Pan L, Hu F, Hu Y. Advanced landfill leachate biochemical effluent treatment using Fe-Mn/AC activates O_3/

$Na_2S_2O_8$ process optimization, wastewater quality analysis and activator characterization. Environmental Science and Pollution Research. 2020;**27**: 15337-15349

[54] Xi BD, Wei ZM, Zhao Y, Yang TX. Study on fluorescence characteristic of dissolved organic matter from municipal solid waste landfill leachate. Spectroscopy and Spectral Analysis. 2008;**28**:2605-2608

[55] Wang F, Luo Y, Ran G, Li Q. Sequential coagulation and FeO-O_3/H_2O_2 process for removing recalcitrant organics from semi-aerobic aged refuse biofilter leachate: Treatment efficiency and degradation mechanism. Science Total Environment. 2020;**699**:134371

[56] He XS, Xi BD, Li X, Pan HW, An D, Bai SG, et al. Fluorescence excitation-emission matrix spectra coupled with parallel factor and regional integration analysis to characterize organic matter humification. Chemosphere. 2013;**93**: 2208-2215

Section 2

Novel Fluorophores and Optical Chemosensors

Chapter 3

Fluorescence Imaging Enhanced by Members of the Graphene Family: A Review

Hu Li and Raffaello Papadakis

Abstract

Graphene is a two-dimensional allotrope of carbon with a range of highly attractive physicochemical properties suitable for a wide variety of applications. In the context of fluorescence imaging graphene and its derivatives have recently started to gain more attention since they could assist in the enhancement of imaging of cells, tissue, or other biologically relevant samples such as cell organoids for example mitochondria as well as in the imaging of cancer cells, tumors, and various pathogens. This chapter attempts to cover the most relevant, recent advances in this growing research field. Some basic information on the physical and (photo)chemical properties of important members of the graphene family is provided. Additionally, novel approaches involving graphene-based materials (GBMs) in cellular and tissue imaging systems are reviewed. Important examples of contemporary applications of GBMs in cancer detection using fluorescence imaging are also presented. The specific role of graphene (or other GBMs) in each case is explained and analyzed. Finally, future perspectives and novel applications of fluorescent imaging techniques involving GBMs are discussed.

Keywords: graphene, fluorescence, imaging, 2D-materials, microscopy, cellular imaging, cancer detection

1. Introduction

1.1 Graphene-based fluorescence imaging applications

Fluorescence imaging is a non-invasive approach that utilizes fluorescent probes to generate photons and provides more sensitivity, specificity, and less harm than other imaging methods [1]. It can be used to monitor cells, tissues, and living organisms *in situ* and analyze specific biomolecules. This is why fluorescence imaging is considered as one of the most essential tools for biomedical research [2]. The fluorescence imaging probes should not be cytotoxic while they should exhibit adequate resistance to photobleaching caused by the natural features of the biological system. In this regard, graphene-based nanomaterials are presently considered as viable alternatives for fluorescence imaging. However, because graphene is a zero-bandgap material, fluorescence cannot be detected in pristine graphene [3]. Conveniently, numerous studies have successfully fragmented or functionalized 2D graphene into 0D graphene quantum dots (GQDs) or carbon nanodots (CNDs) [4]. GQDs consist of graphene lattices and are commonly employed in fluorescence imaging due to quantum confinement and edge effects. Not only do they exhibit the distinct physical and chemical characteristics of conventional graphene, but they also display excellent biocompatibility, low cytotoxicity, and easy functionalization opportunities [5]. However, the poor fluorescent quantum yields often encountered along with the non-specificity of GQDs restrict their wide application in fluorescence imaging. Various research attempts have demonstrated that surface modification of GQDs can effectively alter their chemical activity, electrical structure, and quantum yield, consequently improving their photoluminescence capabilities [6, 7]. Since the realization of nitrogen doped GQDs [8], heteroatomic doping of GQDs has proven to be an effective technique for increasing quantum yield and tuning photoluminescence wavelength. For instance, in Wang et al. utilized the solvothermal method to synthesize boron-doped GQDs (B-GQDs) and phosphorus-doped GQDs (P-GQDs) [9]. B-GQDs and P-GQDs have emission wavelengths of 460 and 630 nm, respectively. Due to the matched band structure, rapid energy transfer between the blue-emitting B-GQDs and the orange-emitting P-GQDs can result in efficient fluorescence emission in the P-GQDs when the blue-emitting B-GQDs are excited at the ideal excitation wavelength of 460 nm. Furthermore, with effective energy transfer, the quantum yield of P-GQDs improves to 0.48, which is significantly greater than the quantum yield of pure P-GQDs. Similarly In Gong et al. synthesized nitrogen and bromine co-doped GQDs (NBr-GQDs) that used a facile deflagration approach for the first time. The quantum yield of NBr-GQDs was up to 52%, with low cytotoxicity, great pH resistance, and stable photoluminescence intensity [10]. Theoretical calculations indicate that N and Br co-doping may reduce the band gap between excited singlet states, significantly improving the photoluminescence performance of GQDs. Recently, Li et al. have also synthesized manganese and boron-nitrogen-doped graphene quantum dots (Mn-BN-GQDs) using the hydrothermal synthesis method and employed them in biosensors [11]. The results reveal that Mn-BN-GQDs have excellent fluorescence characteristics and quantum yield, low cytotoxicity, and high biocompatibility, indicating a promising future for the advancement of bioimaging. On the other hand, graphene oxide (GO), one of the most significant graphene derivatives, has a heterogeneous electrical structure that enables it to emit fluorescence in a particular wavelength range. GO exhibits exceptional features including great mechanical strength, strong photostability, simple surface modification, and photoluminescence that is wavelength dependent. The presence of epoxide functional groups on its surface offers an easily chemically modifiable substrate allowing for the conjugation or interaction with a wide range of biomolecules *via* a variety of covalent/non-covalent interactions, electrostatic forces, absorption, and hydrogen bonding [12]. GO mainly consisting of sp2 and sp3 carbons, generate electron and optical band gaps, allowing it to fluoresce over a broad range of wavelengths and serve as a donor for fluorescence resonance transfer. Moreover, GO possesses effective fluorescence quenching characteristics. Most often, GO acts as the energy acceptor, whereas organic dye acts as the donor. Since the donor's emission spectrum overlaps with the acceptor's absorption spectrum, fluorescence resonance transfer can occur between the two, resulting in fluorescence quenching of the fluorescent dye. In particular, the functional group types, localized domains, lateral dimensions, and solvent dopants can significantly impact the electronic energy transition and fluorescence characteristics of GO [13].

Currently, Numerous studies have demonstrated that GO and its derivatives may generate fluorescence signals of various colors. Based on a study by Kalluru, Nano-GO exhibits single-photon excitation wavelength-dependent photoluminescence in the visible and short near-infrared ranges, making it appropriate for multicolor fluorescence imaging *in vivo* [14]. This particular property of GO in a single biological platform makes this nanomaterial a promising candidate for clinical applications in the early diagnosis of a variety of disorders. Wang et al. reported for the first time the metal fluorescence enhancement of GO and discovered that the fluorescence intensity of GO on an Ag substrate was approximately 10 times stronger than that on glass [15]. Furthermore, unlike other fluorescent materials, GO in direct contact with the metal exhibits a high metal fluorescence enhancement without quenching, indicating that GO might be employed as a fluorescent probe for 3D optical imaging and sensing.

The following section provides a comprehensive introduction to the fluorescence imaging mechanism of graphene quantum dots and graphene oxide, as well as some reviews of current achievements in the field of graphene-based fluorescence imaging application. Fundamentally, to increase the performance of graphene and its derivatives in bioimaging applications, it is necessary to properly control their sizes, surface coatings, and components in order to maximize their photoluminescence capabilities. It is anticipated that the unique structure and exceptional capabilities of graphene and its derivatives offer new opportunities for disease detection and clinical treatments, have promising application prospects in fluorescence imaging, and play a crucial role in fostering the growth of the biomedicine industry.

2. GO, GQDs and CNDs in fluorescence imaging: why so important?

2.1 Graphene oxide

Graphene oxide is one of the most commonly used forms of the graphene-based materials (GBMs) family nowadays. This comes as no surprise since it is a very stable multifunctional material with a wide range of superior properties and hence a broad application scope [16]. GO is essentially the oxidized form of graphene which can be produced through a variety of oxidative methods [17–19]. Structurally, GO encompasses a variety of functional groups, the most important being: carboxy, hydroxy, epoxy, and keto- or aldehyde groups. Hydroxy groups can be found both at the edges and the surface of GO, epoxy on the surface areas and the rest mostly at the edges [20]. Their abundance and ratios are sensitive to the preparation method. These functional groups give rise to a quite dipolar character of this GBM and at the same time generate a range of excitation possibilities due to the presence of oxygen as the main heteroatom involving lone pairs of electrons. As a result, transitions such as n-σ^* n-π^* as well as π-π^* are the most important [21]. When it comes to the electronic and emission properties which are of primary interest for applications within fluorescence imaging, GO is known to exhibit a range of very interesting features:

2.1.1 Excitation energy dependent fluorescence

This is a property which is also observed in GQDs. Unlike most of the conventional fluorophores, GO can result in different emission energies (i.e. a variety of colors of emitted light) when photo-excited at different wavelengths (energies) [22]. Specifically, upon increasing excitation wavelength from 325 nm to 650 nm, a red shift of the GO fluorescent band is observed [23]. This effect in terms of applications in fluorescence imaging is very important since by employing a single material it is possible to obtain a wide range of emitted light colors simply employing different laser energies for excitation.

2.1.2 Fluorescence resonance energy transfer (FRET)

FRET occurs when a donor chromophore, being in its electronically excited state, transfers energy to an acceptor chromophore *via* nonradiative dipole–dipole coupling [24]. The efficiency FRET is proportional to the reciprocal sixth power of the distance between the acceptor and the donor rendering this effect very sensitive to the donor-acceptor distance (see **Figure 1**) [25]. A wide variety of donor and acceptor fluorophore combinations [26] have been reported so far and FRET is considered as an eminent tool for biomolecular research and applications thereof [27, 28]. When it comes to GO, it is well established that GO can function as an efficient energy acceptor (EA) causing the quenching of fluorescence of a variety of energy donor organic fluorophores [29]. The benefits of using GO as an EA are multiple; the variety of functionalization possibilities its structure offers, the large number of binding sites as a result of its size, the high internalization to a wide range of cells, and the very effective energy transfer though a long distance [30] are some of them. In practice, the approach of employing GO, a fluorophore acting as the electron donor, and a functionality, factor, or molecule exhibiting specificity to a target analyte is widely used in many sensor and imaging attempts. Biorelevant analytes and biomarkers such as antibodies, microRNA [31, 32], folate receptor [33] as well as many other examples of analytes [29, 34] have been reported so far.

Additionally, GO can also be employed as an energy donor (ED) in FRET systems involving GO and an acceptor fluorophore. Such systems might involve either organic

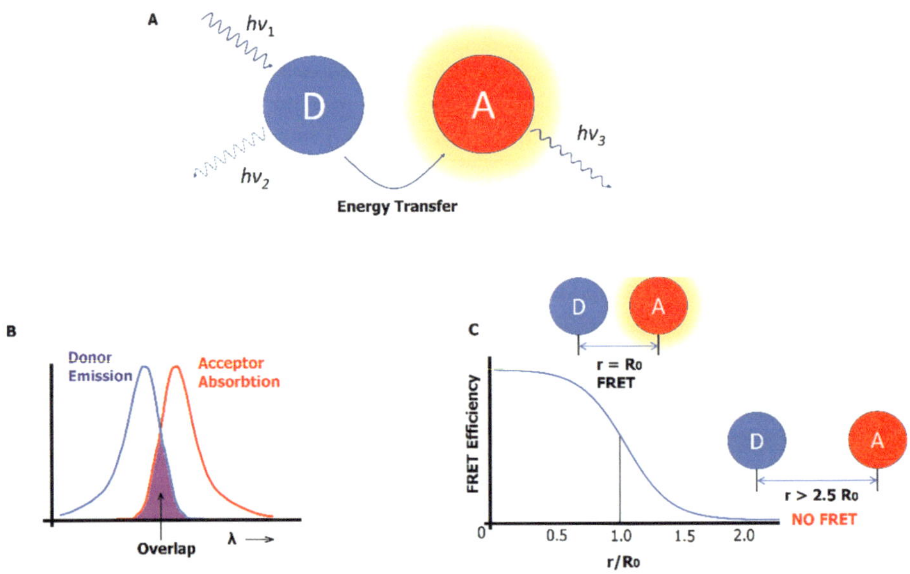

Figure 1.
Fundamental info about FRET. (A) Illustration of the FRET interaction between a donor and an acceptor fluorophore. (B) Example of successfully donor emission/acceptor absorbance overlapping spectra (C) dependence of FRET efficiency on the distance between the donor and the acceptor.

dyes or noble metal nanoparticles acting as EAs corresponding to GO as ED [35, 36]. A variety of bio-relevant sensor applications based on GO/EA FRET systems have been reported so far allowing for the detection of antibodies [35] pharmaceutical screening/imaging, diagnostic tools [36, 37], etc. Some more detailed examples are presented in the next paragraphs.

2.1.3 Fluorescence of GO

The fluorescence band of GO is very broad and thus not suitable for accurate sensor and fluorescence imaging applications. Nonetheless, GO exhibits NIR emission when photoexcited in the NIR region and due to this feature GO is currently finding application in TP microscopy and imaging of biological samples [38, 39]. It is noteworthy that the excitation dependent emission (*vide supra*) allows for use of GO in the two biological wavelength-windows of 1000–1350 nm and 650–950 nm in which biological samples are nearly transparent [40]. This important fact justifies one more important use of GO in biological imaging in addition to FRET applications. Moreover, research attempts to tune the emission colors of GO have been also reported. Mei et al. [41] came out with an oxidative methodology allowing for controlled modifications of GO leading to emissions shifts from brown to cyan with no excitation wavelength alteration. Indeed, by merely varying oxidation reaction times of GO nanosheets was proved to result in controllable and accurate tuning of the emission properties. This approach is thought to introduce new opportunities in cellular imaging as well as in multiplex encoding analysis.

2.2 Graphene quantum dots (GQDs) and carbon nanodots (CNDs)

A very interesting family of materials belonging to the wide carbon family is the Carbon dots family. Two important classes of carbon dots are Graphene quantum dots (GQDs) and carbon nanodots (CNDs). GQDs and CNDs are currently in the forefront of research in the field of bioimaging and specifically within the research and development of novel fluorescence imaging materials and techniques. The ease of production of CNDs and GQDs, their high structural versatility and the wide range of structure-properties modulation opportunities, as well as their biocompatibility and bright fluorescence are the main reasons associated with their high relevance for bioimaging. Up to date a broad range of production methods have been published.

Several methods have been reported for producing GQDs and CNDs, with solvothermal, [42] microwave-assisted, [43] and electrochemical methods [44] being the most commonly used. The choice of the synthetic method used relies on the target batch-size and desired structure and properties of the final materials. It is currently established that for their production inert conditions e.g., heating at temperatures as low as even 120°C are adequate. As a comparison earlier methods employed drastic conditions involving laser ablation [45] or even temperatures higher than 900°C [46]. Nonetheless, it is well agreed by many researchers that milder production conditions can indeed result in very interesting properties since many of the microwave-assisted and hydrothermal methods leave parts of the molecules being subjected to heat or microwaves respectively, unaffected. Hence, the surface of the GQDs and CNDs produced by this "mild processing" can encompass a variety of functional groups [47, 48].

Up to date a wide variety of GQDs and CNDs have been synthesized *via* combining simple compounds such as citric acid, glucose or mono- or di-saccharides and urea, formamide or amino acids [46]. Another trend is to employ natural occurring sources such as fruit extracts, plant specimens etc. which upon treatment (usually microwave or hydrothermal) yield in GQDs and CNDs with a large variety of properties [49, 50]. This research field is highly active which promotes more sustainable and inexpensive production strategies for these vey important class of GBMs.

Why are GQDs and CNDs so important for fluorescence imaging is strongly associated to their outstanding photophysical properties. GQDs and CNDs exhibit photoluminescence (fluorescence emission by many authors) of often high quantum yields in a variety of media. The photoluminescence can be highly dependent on the size and even the shape of the dots [51] but mainly on the surface defects that they exhibit. In fact, without the surface passivation of carbon dots the observed photoluminescence is very limited. Post functionalization of their surface and/or incomplete "termination" reactions/partial carbonization during their production can efficiently enhance their photoluminescent properties [46]. Probably the most remarkable of the properties of GQDs and CNDs is the excitation-dependent photoluminescence [46]. This property (which also GO exhibits, *vide supra*) has a number of important implications on the potential uses of these materials. Specifically, the fact that different emission colors can be obtained by the same material just by choosing a suitable excitation wavelength, is highly appreciated in bioimaging where the use of multiple fluorophores can lead to complications and misinterpretations of the observed phenomena. Indeed, problems such as overlapping emission spectra, photobleaching, and phototoxicity are often encountered [52–54]. These adverse effects can be minimized or even eliminated when using carbon dots. The low toxicity, high water solubility, and very high stability, of carbon dots render this class of GBMs a very attractive option for bioimaging [46]. In the following paragraphs the use of GQDs and CNDs in contemporary fluorescent imaging technologies and applications is discussed.

3. Cellular and tissue imaging systems involving graphene-based materials

Contemporary cell-imaging methods have facilitated the advancement of a wide range of biologically relevant assays aiming at a variety of therapeutic fields and revolutionized the R&D relating to drug design. Cellular imaging encompasses the application of a system or technology required for the visualization of a single cell, cell population, or subcellular structure. Even though a wide range of technologies, methodologies, and molecules enabling cell imaging do exist, there is a constant need for the development of novel systems of higher accuracy, fidelity, specificity, low cost, low cytotoxicity, and high photo- and chemical stability. In terms of fluorescent molecular materials, many of these requirements are fulfilled by members of the wide family of GBMs. In this section, the most recent developments falling in this area of research and technology are reported.

The fluorescence imaging of biomolecules particularly proteins and DNA is an important field of research and technology that can enable the visualization of proteins in cells and tissues with the use of fluorescent probes. Its importance is rationalized as high in terms of the various opportunities for the studies of localization and dynamics of proteins in living cells and tissues that it can offer (e.g. studies of protein–protein interactions, protein folding, and protein degradation) [55–58].

Indeed, graphene and other GBMs have been shown to act as useful platforms for fluorescent sensing of biomolecules including DNA and a variety of proteins and this opens a variety of opportunities in cell and tissue fluorescent imaging [59–62].

Bovine serum albumin (BSA) is a globular protein of animal origin (cow) that is used in a plethora of biochemical applications due to its stability and lack of interference with biological reactions [63]. Kuchlyan et al. performed a thorough study on the interactions of BSA with GO. The group employed a set of spectroscopic methods such as fluorescence correlation spectroscopy (FCS), Fluorescence Lifetime Imaging Microscopy (FILM), and Circular Dichroism (CD). For the study, BSA was labeled with the bright fluorescent Alexa Fluor 488 (AF488). They concluded that GO exhibits a pronouncedly strong interaction with BSA. GO was proved to have a drastic fluorescent quenching effect on AF488-BSA [64]. On the other hand, Yang et al. recently reported on the advancement of a highly sensitive nanosystem based on GO corresponding to microRNA (miRNA) which can be applied in living cells as well as *in vivo* [65]. GO acted as an efficient quenching agent against a molecular beacon labeled with the bright fluorophore cyanine-5 (Cy5). In the presence of the specifically targeted analyte miRNA, fluorescence was recovered allowing detection in cells or tumor tissue samples at very low levels [65].

Very recently Reagen et al. developed a novel class of GQDs exhibiting near-infrared (NIR) fluorescence (emission centered at λ = 860 nm) derived from biomass obtained from organic source and prepared through pyrolysis. The prepared GQDs were tested for cell imaging in two distinct cell lines namely RAW 246.7 (macrophage cells) and MCF-7 breast cancer cell line. The results indicated low cytotoxicity as well as substantial internalization through endocytosis. Moreover, the GQDs exhibited a marked aptitude in detecting Hg^{2+} ions in biological samples enabling NIR fluorescence imaging in cells and toxic heavy metal detection *in vivo* [66]. Such multifunctional GQD-fluorescence imaging systems are currently sought-after. The multifunctional character and multi-use feature of GQDs in cellular imaging has been recently pointed out by a range of research groups [67–69].

Two-photon (TP) microscopy is a fluorescence imaging technique that is particularly well-suited to image-scattering living tissue of up to about 1 millimeter in thickness. It works by shining an intense beam of near-infrared light onto a single point within a sample, inducing simultaneous absorption of two photons at the focal point, where the intensity is the highest [70]. TP microscopy has found nowadays huge applicability in bioscience. Nonetheless, a typical restriction that TP- microscopy/bioimaging techniques exhibit is that they rely on single-color fluorescence changes. Due to the special emission properties of GQDs among a range of other beneficiary properties (*vide supra* paragraph 2 for details) these nanomaterials have become very attractive for fluorescence imaging. Zhao et al. as early as 2016, reported the development of a dual-emission and TP GQD suitable for imaging applications targeting hydrogen peroxide (H_2O_2) as an analyte. The fluorescence response of the TP-GQD towards H_2O_2 was shown to be fast and very specific and renders the mapping of the production of endogenous H_2O_2 in living cells and deep tissues feasible. This probably constitutes the first published example pertaining to a dual-emission, TP-GQD of high specificity and applicability in cell and tissue fluorescence imaging [71].

In Wang et al. reported on the development of some GQDs through a hydrothermal method utilizing 1,3,6-trinitropyrene and $(NH_4)_2SO_3$. The resulting GQDs encompassing amino as well as sulfate groups were evaluated in terms of their TP fluorescence efficiency in the context of cellular imaging. A very high TP absorption cross-section was observed and evaluated as significantly higher as compared to

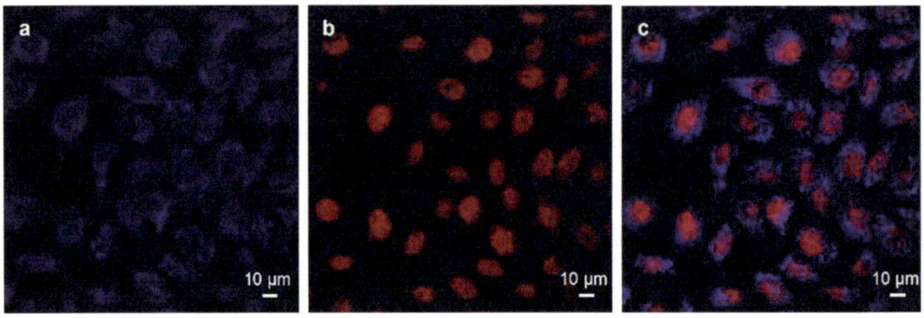

Figure 2.
(a) Image of GQDs at 405 nm excitation. (b) Image of nuclear view red dye at 633 nm excitation. (c) Merged image (obtained with permission from supplementary information of Ref. [72]).

traditional/conventional fluorophores. The research group further performed tests in a cell line (HeLa cells) and found out that the GQDs were internalized in the cytoplasm providing very bright and clear cell images (see **Figure 2**) [72].

Earlier Sapkota et al. reported the synthesis of GQDs of tunable size and explored their capacity in fluorescence imaging. It was found that GQDs with a size between 15 and 35 nm exhibit vivid fluorescence (quantum yields of 0.64 by average) as well as high TP absorption (TPA) cross sections, which renders these GQDs excellent candidates for fluorescence imaging. Indeed, their use in cellular imaging was evaluated on living epithelial cells and even though internalization was observed, entering the nucleus was not possible [73].

Chen et al. as early as in 2015 advanced an aptameric sensor with nano dimensions based on graphene which is capable of inducing/enhancing the fluorescence activation imaging of cytochrome c (Cyt c; a major mediator in cell apoptosis released from mitochondria) [74]. In order to achieve this, Chen et al. connected a fluorophore-tagged DNA aptamer on graphene nanosheets modified with PEG polymer chains. The fluorescence of the fluorophore is inhibited due to the presence of graphene. Yet, dissociation of the a fluorophore-tagged DNA aptamer from graphene occurring immediately after cytosolic release of Cyt C, triggers the fluorescence and empowers real-time visualization of the Cyt c release kinetics. This nanosensing technology is envisioned to exhibit potential applications in visualization of key molecular factors in apoptotic signaling which are critical for cell biology and clinical theranostics.

Wang et al. on the other hand developed an innovative nanosensor employing graphene quantum dots (GQDs) which were conjugated to gold nanoparticles (AuNPs). The nanosensor was found to efficiently serve as a sensor of endogenous biological cyanide ions. This graphene-based nanomaterial further exhibits efficient TP excitation and exploits the drastic quenching efficiency of AuNPs and thus it can accomplish detection cyanide limits as low as 0.52 μM. This, combined with the potential of deep penetration depth of approx. 400 μm render this nanomaterial a perfect candidate for tissue imaging of cyanides [75].

Similarly, Hong et al. also employed a combination of AuNPs and a GBM (GO) for the development of fluorescent imaging/sensing system with specific applicability in monitoring intracellular telomerase activity. Their fluorescence imaging is applicable to a variety of living cells and it was tested toward the aptitude to distinguish normal-form cancer cells [76].

One of the pioneering works reporting GQDs-based nanosystems utilized in TP-induced fluorescence imaging was that by Liu et al. published as early as 2013. The research group produced some N-doped GQDs employing a solvothermal method in which dimethylformamide (DMF) was used both as solvent as well as N-source. The reported GQDs exhibit a marked TP absorption cross-section of nearly 48 kGM (GM: Göppert Mayer units) and were proved to function well even at imaging depth in tissue samples as large as 1800 μm. The development of these GQDs has been an early example of the immense potentials of GQDs in fluorescence bioimaging with applications in the broad biological and biomedical research fields [77].

4. GBMs in cancer detection through fluorescence imaging

Cancer currently constitutes one of the major mortality causes for humans. Ensuring enhanced cancer therapies requires improvement of cancer diagnostic techniques. In recent years, GBMs have found applicability in cancer detection. The wide range of suitable properties and attributes of this broad range of materials are considered for efficient cancer imaging.

Campbell et al. recently reported on the use of GQDs involving amino, hydroxy as well as carboxy functional groups and doped with nitrogen, nitrogen and boron or sulfur, in spectrally distinguishing among healthy and various types of cancer cells. In addition to this, the authors evaluated the pH-responsiveness of the investigated GQDs exploiting their wide range of emitted light wavelengths spanning from the visible to near-infraredred (NIR) part of the electromagnetic spectrum (**Figure 3**) [78].

Wu et al. on the other hand developed a new type of graphene-based nanomaterial functionalized with Anti-EpCAM antibodies and galactose-rhodamine-polyacrylamide

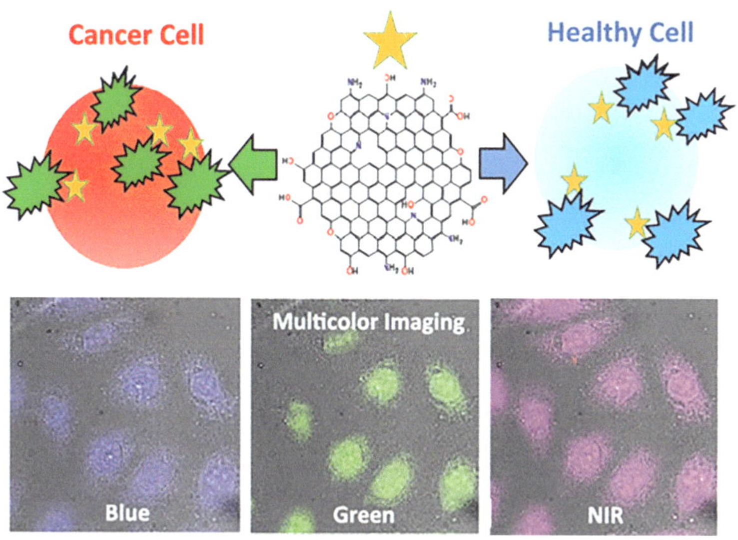

Figure 3.
Illustration depicting the structure of GQDs developed by Campbell et al. and the possibility of distinguishing cancer from healthy cells. Lower images depict the variety of colors emitted by the GQDs as a result of their excitation at different wavelengths (excitation-dependent emission). Obtained with permission from supplementary information of Ref. [78].

nanoparticles with a high aptitude to recognition of hepatocellular carcinoma cells (HCC-CTCs). The presence of graphene is dual here, acting both as a nanoparticle carrier platform as well as a strong quencher of rhodamine's fluorescence. Upon capturing and endocytosis of the aforementioned nanoparticles, fluorescence is recovered and the fluorescence imaging of HCC-CTCs can thus be efficiently achieved (**Figure 4**) [79].

It is important to mention that the aforementioned strategy involving graphene as a quencher of fluorophores connected to agents that can efficiently/specifically bind cancer cells or other bio-targets, has been repeatedly proposed within the context of fluorescence bio-imaging. A variety of recently published attempts towards bioimaging targets of biologically relevant analytes such as DNA [80], ions, [81] antibiotics [82], etc. have recently been reviewed [83].

Nurannabi et al. as early as 2014, developed GQDs bearing -OH and -COOH functionalities with an average size of 5 nm exhibiting significant red photoluminescence when excited at 655 nm (**Figure 5**). Their photoluminescence behavior was shown to enhance visualization of deep tumor tissues in experimental animals. Moreover, the described GQDs were tested *in vivo* as photodynamic therapeutic agents against MDA-MB231 cancer cells and it was proved that they exhibit a marked phototherapeutic activity [84].

In Pramanik et al. reported on the use of aptamer-conjugated GO in TP imaging (TPI) of breast cancer cells (specifically of SK-BR-3 cells). The developed GBM displays a drastic 2-photon absorption and marked photostability after even long irradiation sessions [38].

In Narasimhan et al. reported on the use of GQDs produced through laser ablation for use in both *in vitro* and *in vivo* fluorescence imaging of MCF-7 breast cancer cells. The reported GQDs were bearing hydroxyl and carboxylic acid groups at the edges as well as on their surface and it was observed that uptake of the GQDs by the

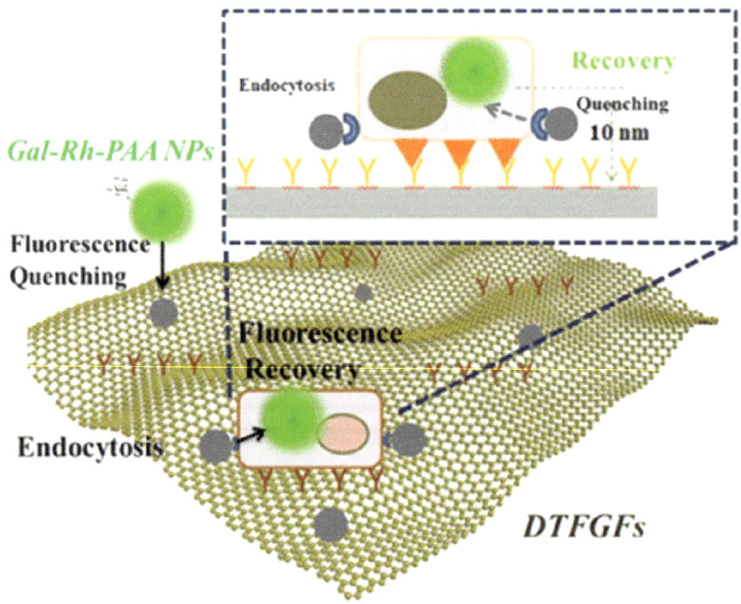

Figure 4.
Illustration of graphene-based nanomaterial developed for recognition of hepatocellular carcinoma cells (HCC-CTCs). Obtained with permission from supplementary information of Ref. [79].

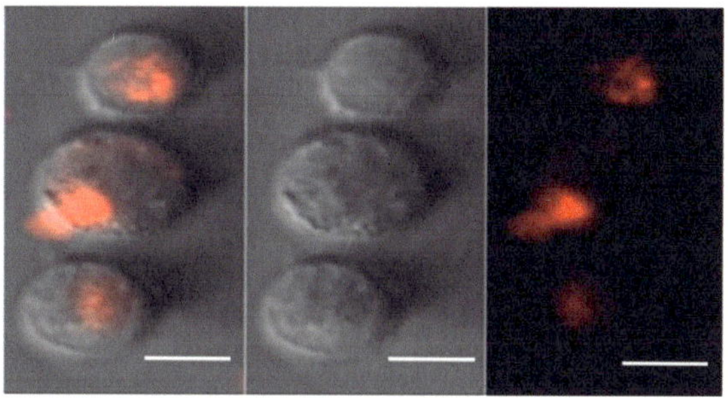

Figure 5.
Confocal laser scanning images of MDA-MB231 cells after treatment of cGdots (100 μg/mL). Scale bar: 20 μm (obtained with permission from supplementary information of Ref. [84]).

MCF-7 cells was feasible through endocytosis rendering imaging of implanted cells in mice possible through an intense red fluorescence observed upon excitation at 610 nm [85].

In Liu et al. proposed the use of graphitic carbon nitride nanosheets as a scaffold allowing for the detection of hyaluronase (HAase). Similarly to the already described strategies for cancer cell/tissue imaging, a cancer marker is targeted as an analyte (in this particular case HAase). Imaging of HAase in cancer tissues can be achieved through the activable two-photon fluorescence of the developed graphitic carbon nitride nanosheets in presence of HAase [86].

Some years earlier Park and coworkers utilized hyaluronic acid (HA) instead of HA-ase to accomplish efficient target-specific delivery of GQDs. In their strategy, the researchers tethered HA to GQDs and thus synthesized brightly fluorescent nanoparticles with an approximate size of 20 nm [87]. These interesting HA-GQD conjugates were tested and it was found that they exhibit strong fluorescence in CD44 overexpressing A549 cells as well as in *in vivo* experiments involving CD44 receptor overexpressing tumor-bearing balb/c female mice. CD44 antigen is a cell-surface glycoprotein involved in cell–cell interactions, cell adhesion and migration. It is the receptor for hyaluronic acid and can also interact with other ligands e.g. matrix metalloproteinases (MMPs) etc. [87]. Its specificity to HA explains the very good emission response of the HA-GQDs by Park and coworkers. As a step further the efficiency of HA-GQDs to specifically curry chemotherapeutic drugs such as doxorubicin to cancer cells was evaluated as high rendering this smart nanomaterial both a good nanosensor and a drug nanocarrier.

In a similar fashion, Goreham et al. investigated the role of GO modification with folic acid in fluorescence lifetime imaging of HaCaT cells. Since folate receptor is an important recognized biomarker [88] currently considered for new diagnostic tools for cancer, modification of graphene oxide with folic acid was considered in this study. The water solubility green fluorescence (upon photoexcitation at 305 nm) and the evaluated low toxicity of GO indicated the high potentials of these graphene family members in cancer fluorescence imaging and corresponding diagnostics [89].

Liu et al. also utilized GO as a perfect platform on which a molecular beacon (a type of oligonucleotide hybridization probe which is capable of specifically detecting the presence of nucleic acids) having a couple of Cyanine-5 (Cy-5) fluorophore units at its both ends. The fluorescence of Cy-5 was quenched both due to self-quenching and fluorescence resonance energy transfer (FRET) caused by GO. With this nanosystem, the researchers achieved detection of microRNAs even at concentrations a low as 30 pM. This nanoimaging technique was found to be applicable to a variety of cancer tissues and cells [65].

Additionally, Kumawat et al. reported the use of GQDs obtained *via* a green chemistry approach in fluorescence nuclear imaging in a variety of cell lines. It was observed that the reported GQDs even deprived of further support of external targeting agents (such as folic acid, hyaluronic acid etc.) exhibited a drastic propensity to self-localize into cell nuclei [90].

Moreover, Fan et al. investigated the use of some pH-responsive GQDs (pRF-GQDs) which were synthesized *via* an electrochemical method. The pRF-GQDs were found to undergo a fluorescence color transition between green and blue by varying the pH and developed a method which enable distinguishing between healthy and cancer tissues [91]. These are two important recent examples of the immense potentials of GQDs in bioimaging of cancer and cancer diagnosis advancements.

While fluorescence can be readily used in the detection and imaging of a variety of tumor cells, Magnetic resonance imaging (MRI) techniques enhanced through the use of GQDs appear to be very attractive due to the ease of production, size, easy structure modulation as well as multi-purpose character of GQDs [92]. Zhang et al. nearly a decade ago reported the use of GO-gadolinium (GO-Gd) complexes for the enhancement of and quality improvement of the MR-imaging of cancer. They showed that the developed GO-based material not only can operate as an enhancing agent for MRI but furthermore serves as a material for fluorescent imaging with anticancer-drug delivery aptitude (**Figure 6**). This early finding clearly showcases the immense possibilities and versatility of GBMs in bioimaging [93].

More recently and in a similar fashion, Yang et al. reported on the use of surface-modified GQDs by polyethylene glycol (PEG) which were functionalized with the Gd-DOTA complex (where DOTA stands for tetraazacyclododecanetetraacetic acid) in cancer imaging through MRI and fluorescence imaging. The research group managed to significantly increase the relaxivity by regulating the length of the PEG linkers and hence advanced a novel MR contrast agent with immense potentials within cancer-imaging [94].

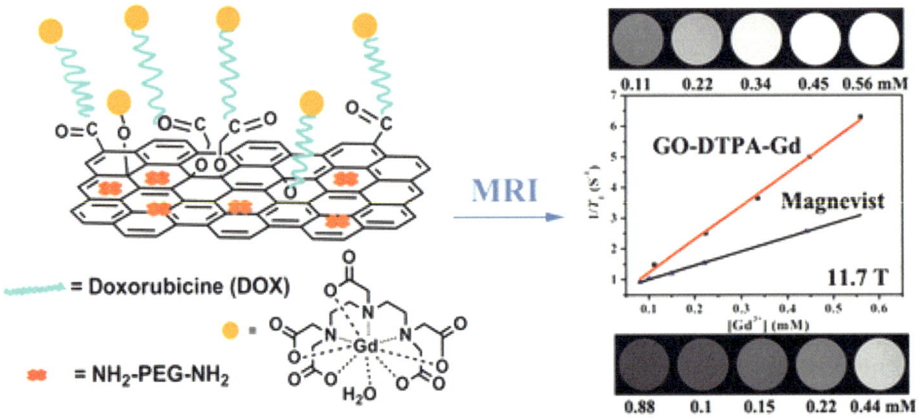

Figure 6.
Illustration of the Gd-bearing GO-based platform developed by Zhang et al. allowing for enhanced MR-cancer imaging as well as for targeted drug delivery applications. Obtained with permission from supplementary information of Ref. [93]).

5. Conclusions and future perspectives and uses of fluorescent imaging involving GBMs

There is a wide range of current and potential future applications of fluorescence imaging, two of the most promising being fluorescence-guided surgery (FGS) [95] and robotic-assisted fluorescence surgery (RAFS) [96]. The applications are numerous while probably the most relevant ones lie in the field of surgical oncology [95, 97]. The idea of using tumor-targeted imaging agents in the context of this developing research and technology area is considered as a promising strategy for intraoperative cancer detection. In this regard, the broad family of GBMs and their attributes can demonstrate an important role as potent fluorescent/emissive materials of low toxicity, biocompatibility and tunable optical properties [46]. A variety of GQDs have been reported so far labelling tumor cells and tissue thus potentially facilitating tumor surgery [91, 98, 99]. Particular interest has been placed on the use on pH-responsive fluorescent GQDs (pRF-GQDs) as transition in the emitted light color depending on the pH can help distinguish healthy tissue from tumors (extracellular microenvironment of tumors exhibits lower pH) [33, 91]. Based on the so-far published research works, it becomes apparent that the multifunctional character of GBMs could facilitate tumor FGS in the future and in general bioimaging and the detection of disease [100, 101].

Acknowledgements

The authors are grateful to Dr. K. Tzavlaki for fruitful discussions on bioimaging.

Conflict of interest

The authors declare no conflict of interest.

Acronyms and abbreviations

AuNPs	gold nanoparticles
BSA	bovine serum albumin
CNDs	carbon nanodots
CD	circular dichroism
CD44	antigen is a cell-surface glycoprotein
Cyt c	cytochrome c 1
DMF	dimethylformamide
EA	energy acceptor
ED	energy donor
FGS	fluorescence guided surgery
FRET	fluorescence resonance energy transfer
FCS	fluorescence correlation spectroscopy
FILM	fluorescence lifetime imaging microscopy
GBMs	graphene based materials
GM	Göppert Mayer units
GO	graphene oxide

GQDs	graphene quantum dots
HA	hyaluronic acid
HAase	hyaluronase
HaCaT	a spontaneously transformed aneuploid immortal keratinocyte cell line
HCC-CTCs	hepatocellular carcinoma cells
MRI	magnetic resonance imaging
NIR	near infrared
RAFS	robotic-assisted fluorescence surgery
pRF-GQDs	pH-responsive fluorescent graphene quantum dots
PEG	polyethylene glycol
SK-BR-3	a breast cancer cell line
TP	two photon
TPA	two photon absorption

Author details

Hu Li[1,2] and Raffaello Papadakis[3]*

1 Shandong Technology Centre of Nanodevices and Integration, School of Microelectronics, Shandong University, Jinan, China

2 Department of Materials Science and Engineering, Uppsala University, Uppsala, Sweden

3 TdB Labs, Uppsala, Sweden

*Address all correspondence to: rafpapadakis@gmail.com

References

[1] Baker SN, Baker GA. Luminescent carbon nanodots: Emergent nanolights. Angewandte Chemie International Edition. 2010;**49**:6726-6744. DOI: 10.1002/anie.200906623

[2] Zhu S, Meng Q, Wang L, Zhang J, Song Y, Jin H, et al. Highly photoluminescent carbon dots for multicolor patterning, sensors, and bioimaging. Angewandte Chemie International Edition. 2013;**52**:3953-3957. DOI: 10.1002/anie.201300519

[3] Kavitha MK, Jaiswal M. Graphene: A review of optical properties and photonic applications. Asian Journal of Physics. 2016;**25**:809-831

[4] Xin G, Meng Y, Ma Y, Ho D, Kim N, Cho SM, et al. Tunable photoluminescence of graphene oxide from near-ultraviolet to blue. Materials Letters. 2012;**74**:71-73. DOI: 10.1016/j.matlet.2012.01.047

[5] Li L, Wu G, Yang G, Peng J, Zhao J, Zhu J-J. Focusing on luminescent graphene quantum dots: Current status and future perspectives. Nanoscale. 2013;**5**:4015-4039. DOI: 10.1039/C3NR33849E

[6] Li H, Han S, Lyu B, Hong T, Zhi S, Xu L, et al. Tunable light emission from carbon dots by controlling surface defects. Chinese Chemical Letters. 2021;**32**:2887-2892. DOI: 10.1016/j.cclet.2021.03.051

[7] Ai L, Shi R, Yang J, Zhang K, Zhang T, Lu S. Efficient combination of G-C3N4 and CDs for enhanced photocatalytic performance: A review of synthesis, strategies, and applications. Small. 2021;**17**:2007523. DOI: 10.1002/smll.202007523

[8] Li Y, Zhao Y, Cheng H, Hu Y, Shi G, Dai L, et al. Nitrogen-doped graphene quantum dots with oxygen-rich functional groups. Journal of the American Chemical Society. 2012;**134**:15-18. DOI: 10.1021/ja206030c

[9] Wang G, He P, Xu A, Guo Q, Li J, Wang Z, et al. Promising fast energy transfer system between graphene quantum dots and the application in fluorescent bioimaging. Langmuir. 2019;**35**:760-766. DOI: 10.1021/acs.langmuir.8b03739

[10] Gong L, Sun J, Zheng P, Liu Y, Yang G. Yellow fluorescent nitrogen and bromine Co-doped graphene quantum dots for bioimaging. ACS Applied Nano Materials. 2021;**4**:8564-8571. DOI: 10.1021/acsanm.1c02004

[11] Li B, Xiao X, Hu M, Wang Y, Wang Y, Yan X, et al. Mn, B, N co-doped graphene quantum dots for fluorescence sensing and biological imaging. Arabian Journal of Chemistry. 2022;**15**:103856. DOI: 10.1016/j.arabjc.2022.103856

[12] Aliyev E, Filiz V, Khan MM, Lee YJ, Abetz C, Abetz V. Structural characterization of graphene oxide: Surface functional groups and fractionated oxidative debris. Nanomaterials. 2019;**9**:1180

[13] Lai S, Jin Y, Shi L, Zhou R, Zhou Y, An D. Mechanisms behind excitation- and concentration-dependent multicolor photoluminescence in graphene quantum dots. Nanoscale. 2020;**12**:591-601. DOI: 10.1039/C9NR08461D

[14] Kalluru P, Vankayala R, Chiang C-S, Hwang KC. Nano-graphene oxide-mediated In vivo fluorescence imaging and bimodal photodynamic

and photothermal destruction of tumors. Biomaterials. 2016;**95**:1-10. DOI: 10.1016/j.biomaterials.2016.04.006

[15] Wang Z, Zheng L, Cheng Q, Li X, Huang L, Lu Y. Metal-enhanced fluorescence of graphene oxide sheets. Analytical and Bioanalytical Chemistry. 2022;**414**:3625-3630. DOI: 10.1007/s00216-022-04001-x

[16] Razaq A, Bibi F, Zheng X, Papadakis R, Jafri SHM, Li H. Review on graphene-, graphene oxide-, reduced graphene oxide-based flexible composites: From fabrication to applications. Materials. 2022;**15**(3):1012. DOI: 10.3390/ma15031012

[17] Marcano DC, Kosynkin DV, Berlin JM, Sinitskii A, Sun Z, Slesarev A, et al. Improved synthesis of graphene oxide. ACS Nano. 2010;**4**(8):4806-4814. DOI: 10.1021/nn1006368

[18] Chen J, Yao B, Li C, Shi G. An improved hummers method for eco-friendly synthesis of graphene oxide. Carbon. 2013;**64**:225-229. DOI: 10.1016/j.carbon.2013.07.055

[19] Sun L. Structure and synthesis of graphene oxide. Chinese Journal of Chemical Engineering. 2019;**27**(10):2251-2260. DOI: 10.1016/j.cjche.2019.05.003

[20] Dimiev AM, Eigler S, editors. Graphene Oxide: Fundamentals and Applications. John Wiley & Sons, Ltd; 2016

[21] Gao W, editor. Graphene Oxide. Reduction Recipes, Spectroscopy, and Applications. Springer Cham; 2015. DOI: 10.1007/978-3-319-15500-5

[22] Li M, Cushing SK, Zhou X, Guoc S, Wu N. Fingerprinting photoluminescence of functional groups in graphene oxide. Journal of Materials Chemistry. 2012;**22**:23374-23379

[23] Zheng P, Prof W, N. Fluorescence and sensing applications of graphene oxide and graphene quantum dots: A review. Chemistry, an Asian Journal. 2017;**12**(18):2343-2353. DOI: 10.1002/asia.201700814

[24] Helms V. Fluorescence resonance energy transfer. In: Principles of Computational Cell Biology. Weinheim: Wiley-VCH; 2008. p. 202

[25] Schneckenburger H. Förster resonance energy transfer–what can we learn and how can we use it? Methods and Applications in Fluorescence. 2019;**8**(1):013001. DOI: 10.1088/2050-6120/ab56e1

[26] Bajar BT, Wang ES, Zhang S, Lin MJC, B. T. A guide to fluorescent protein FRET pairs. Sensors. 2016;**16**(9):1488. DOI: 10.3390/s16091488

[27] Algar WR, Hildebrandt N, Vogel SS, Medintz IL. FRET as a biomolecular research tool — Understanding its potential while avoiding pitfalls. Nature Methods. 2019;**16**:815-829. DOI: 10.1038/s41592-019-0530-8

[28] Kaur A, Kaur P, Ahuja S. Förster resonance energy transfer (FRET) and applications thereof. Analytical Methods. 2020;**12**:5532-5550. DOI: 10.1039/D0AY01961E

[29] Tian F, Lyu J, Shi J, Yang M. Graphene and graphene-like two-denominational materials based fluorescence resonance energy transfer (FRET) assays for biological applications. Biosensors & Bioelectronics. 2017;**89**:123-135. DOI: 10.1016/j.bios.2016.06.046

[30] Prabakaran G, Velmurugan K, David CI, Nandhakumar R. Role of

Förster resonance energy transfer in graphene-based nanomaterials for sensing. Applied Sciences. 2022;**12**(14):6844. DOI: 10.3390/app12146844

[31] Nitu FR, Savu L, Muraru S, Stoian I, Ioniță M. Label-free homogeneous microRNA detection in cell culture medium based on graphene oxide and specific fluorescence quenching. Nanomaterials. 2021;**11**(2):368. DOI: 10.3390/nano11020368

[32] Guo Y, Xu H, Li Y, Wu F, Li Y, Bao Y, et al. Hyaluronic acid and Arg-Gly-asp peptide modified graphene oxide with dual receptor-targeting function for cancer therapy. Journal of Biomaterials Applications. 2017;**32**(1):54-65. DOI: 10.1177/0885328217712110

[33] Battogtokh G, Ko YT. Graphene oxide-incorporated pH-responsive folate-albumin-photosensitizer nanocomplex as image-guided dual therapeutics. Journal of Controlled Release: Official Journal of the Controlled Release Society. 2016;**234**:10-20. DOI: 10.1016/j.jconrel.2016.05.007

[34] Cruz SMA, Girão AF, Gonçalves G, Marques PAAP. Graphene: The missing piece for cancer diagnosis? Sensors. 2016;**16**(1):137. DOI: 10.3390/s16010137

[35] Kim J, Park S-J, Min D-H. Emerging approaches for graphene oxide biosensor. Analytical Chemistry, 2017;**89**(1):232-248. DOI: 10.1021/acs.analchem.6b04248

[36] Kwak S-Y, Yang J-K, Jeon S-J, Kim H-I, Yim J, Kang H, et al. Luminescent graphene oxide with a peptide-quencher complex for optical detection of cell-secreted proteases by a turn-on response. Advanced Functional Materials. 2014;**24**(32):5119-5128. DOI: 10.1002/adfm.201400001

[37] Shi Y, Dai H, Sun Y, Hu J, Ni P, Li Z. Fluorescent sensing of cocaine based on a structure switching aptamer, gold nanoparticles and graphene oxide. The Analyst. 2013;**138**(23):7152-7156. DOI: 10.1039/c3an00897e

[38] Pramanik A, Chavva SR, Fan Z, Sinha SS, Nellore BPV, Ray PC. Extremely high two-photon absorbing graphene oxide for imaging of tumor cells in the second biological window. The Journal of Physical Chemistry Letters. 2014;**5**(12):2150-2154. DOI: 10.1021/jz5009856

[39] Li D, Xue L, Zhu Z, Zhao X, Qian J. Graphene oxide nanoparticles for two-photon fluorescence imaging of zebrafish. Optical and Quantum Electronics. 2016;**48**(11):519. DOI: 10.1007/s11082-016-0783-8

[40] Tsai M-F, Chang S-HG, Cheng F-Y, Shanmugam V, Cheng Y-S, Su C-H, et al. Au nanorod design as light-absorber in the first and second biological near-infrared windows for in vivo photothermal therapy. ACS Nano. 2013;**7**(6):5330-5342. DOI: 10.1021/nn401187c

[41] Mei Q, Chen J, Zhao J, Yang L, Liu B, Liu R, et al. Atomic oxygen tailored graphene oxide nanosheets emissions for multicolor cellular imaging. ACS Applied Materials & Interfaces. 2016;**8**(11):7390-7395. DOI: 10.1021/acsami.6b00791

[42] Huo Y, Xiu S, Meng L-Y, Quan B. Solvothermal synthesis and applications of micro/nano carbons: A review. Chemical Engineering Journal. 2023;**451**(138572):138572. DOI: 10.1016/j.cej.2022.138572

[43] De Medeiros TV, Manioudakis J, Noun F, Macairan J-R, Victoria F, Naccache R. Microwave-assisted

synthesis of carbon dots and their applications. Journal of Materials Chemistry C. 2019;**7**(24):7175-7195. DOI: 10.1039/c9tc01640f

[44] Rocco D, Moldoveanu VG, Feroci M, Bortolami M, Vetica F. Electrochemical synthesis of carbon quantum dots. ChemElectroChem. 2023;**10**(3):e2022011. DOI: 10.1002/celc.202201104

[45] Sun Y-P, Zhou B, Lin Y, Wang W, Fernando KAS, Pathak P, et al. Quantum-sized carbon dots for bright and colorful photoluminescence. Journal of the American Chemical Society. 2006;**128**(24):7756-7757. DOI: 10.1021/ja062677d

[46] Jelinek R. Carbon Quantum Dots: Synthesis, Properties and Applications. 1st ed. Switzerland: Springer International Publishing; 2016

[47] Bhunia SK, Maity AR, Nandi S, Stepensky D, Jelinek R. Imaging cancer cells expressing the folate receptor with carbon dots produced from folic acid. Chembiochem: A European Journal of Chemical Biology. 2016;**17**(7):614-619. DOI: 10.1002/cbic.201500694

[48] Tang L, Ji R, Cao X, Lin J, Jiang H, Li X, et al. Deep ultraviolet photoluminescence of water-soluble self-passivated graphene quantum dots. ACS Nano. 2012;**6**(6):5102-5110. DOI: 10.1021/nn300760g

[49] Chahal S, Macairan J-R, Yousefi N, Tufenkji N, Naccache R. Green synthesis of carbon dots and their applications. RSC Advances. 2021;**11**(41):25354-25363. DOI: 10.1039/d1ra04718c

[50] Hill S, Galan MC. Fluorescent carbon dots from mono- and polysaccharides: Synthesis, properties and applications. Beilstein Journal of Organic Chemistry. 2017;**13**:675-693. DOI: 10.3762/bjoc.13.67

[51] Kim S, Hwang SW, Kim M-K, Shin DY, Shin DH, Kim CO, et al. Anomalous behaviors of visible luminescence from graphene quantum dots: Interplay between size and shape. ACS Nano. 2012;**6**(9):8203-8208. DOI: 10.1021/nn302878r

[52] Zhang Y, Zheng Y, Tomassini A, Singh AK, Raymo FM. Photoactivatable fluorophores for bioimaging applications. ACS Applied Optical Materials. 2023;**1**(3):640-651. DOI: 10.1021/acsaom.3c00025

[53] Qiu X, Xu J, Cardoso Dos Santos M, Hildebrandt N. Multiplexed biosensing and bioimaging using lanthanide-based time-gated Förster resonance energy transfer. Accounts of Chemical Research. 2022;**55**(4):551-564. DOI: 10.1021/acs.accounts.1c00691

[54] Pandey S, Bodas D. High-quality quantum dots for multiplexed bioimaging: A critical review. Advances in Colloid and Interface Science. 2020;**278**:102137. DOI: 10.1016/j.cis.2020.102137

[55] Zhu H, Hamachi I. Fluorescence imaging of drug target proteins using chemical probes. Journal of Pharmaceutical Analysis. 2020;**10**(5):426-433. DOI: 10.1016/j.jpha.2020.05.013

[56] Berezin MY, Achilefu S. Fluorescence lifetime measurements and biological imaging. Chemical Reviews. 2010;**110**(5):2641-2684. DOI: 10.1021/cr900343z

[57] Yu L, Lei Y, Ma Y, Liu M, Zheng J, Dan D, et al. A comprehensive review of fluorescence correlation spectroscopy. Frontiers in Physics. 2021:**9**. DOI: 10.3389/fphy.2021.644450

[58] Terai T, Nagano T. Small-molecule fluorophores and fluorescent probes for

bioimaging. Pflugers Archiv: European Journal of Physiology. 2013;**465**(3):347-359. DOI: 10.1007/s00424-013-1234-z

[59] Lu C-H, Yang H-H, Zhu C-L, Chen X, Chen G-N. A graphene platform for sensing biomolecules. Angewandte Chemie (Int. Ed.). 2009;**48**(26):4785-4787. DOI: 10.1002/anie.200901479

[60] Coreas R, Castillo C, Li Z, Yan D, Gao Z, Chen J, et al. Biological impacts of reduced graphene oxide affected by protein corona formation. Chemical Research in Toxicology. 2022;**35**(7):1244-1256. DOI: 10.1021/acs.chemrestox.2c00042

[61] Yildiz G, Bolton-Warberg M, Awaja F. Graphene and graphene oxide for bio-sensing: General properties and the effects of graphene ripples. Acta Biomaterialia. 2021;**131**:62-79. DOI: 10.1016/j.actbio.2021.06.047

[62] de Lázaro I, Vranic S, Marson D, Rodrigues AF, Buggio M, Esteban-Arranz A, et al. Graphene oxide as a 2D platform for complexation and intracellular delivery of siRNA. Nanoscale. 2019;**11**(29):13863-13877. DOI: 10.1039/c9nr02301a

[63] Bovine serum albumin. Meyler's Side Effects of Drugs. Amsterdam, Netherlands: Elsevier; 2016. p. 1045

[64] Kuchlyan J, Kundu N, Banik D, Roy A, Sarkar N. Spectroscopy and fluorescence lifetime imaging microscopy to probe the interaction of bovine serum albumin with graphene oxide. Langmuir: The ACS Journal of Surfaces and Colloids. 2015;**31**(51):13793-13801. DOI: 10.1021/acs.langmuir.5b03648

[65] Yang L, Liu B, Wang M, Li J, Pan W, Gao X, et al. A highly sensitive strategy for fluorescence imaging of MicroRNA in living cells and in vivo based on graphene oxide-enhanced signal molecules quenching of molecular beacon. ACS Applied Materials & Interfaces. 2018;**10**(8):6982-6990. DOI: 10.1021/acsami.7b19284

[66] Reagen S, Wu Y, Liu X, Shahni R, Bogenschuetz J, Wu X, et al. Synthesis of highly near-infrared fluorescent graphene quantum dots using biomass-derived materials for in vitro cell imaging and metal ion detection. ACS Applied Materials & Interfaces. 2021;**13**(37):43952-43962. DOI: 10.1021/acsami.1c10533

[67] Dutta Chowdhury A, Ganganboina AB, Tsai Y-C, Chiu H-C, Doong R-A. Multifunctional GQDs-concanavalin a@Fe3O4 nanocomposites for cancer cells detection and targeted drug delivery. Analytica Chimica Acta. 2018;**1027**:109-120. DOI: 10.1016/j.aca.2018.04.029

[68] Thakur M, Kumawat MK, Srivastava R. Multifunctional graphene quantum dots for combined photothermal and photodynamic therapy coupled with cancer cell tracking applications. RSC Advances. 2017;7(9):5251-5261. DOI: 10.1039/c6ra25976f

[69] Yang Y, Tang S, Chen D, Wang C, Gu B, Li X, et al. Multifunctional red-emission graphene quantum dots with tunable light emissions for trace water sensing, WLEDs and information encryption. Colloids and Surfaces. A, Physicochemical and Engineering Aspects. 2021;**622**(126593):126593. DOI: 10.1016/j.colsurfa.2021.126593

[70] Suan D, Hampton HR, Tomura M, Kanagawa O, Chtanova T, Phan TG. Optimizing fluorescence excitation and detection for intravital two-photon microscopy. Methods in Cell Biology.

2013;**113**:311-323. DOI: 10.1016/B978-0-12-407239-8.00014-8

[71] Zhao W, Li Y, Yang S, Chen Y, Zheng J, Liu C, et al. Target-activated modulation of dual-color and two-photon fluorescence of graphene quantum dots for in vivo imaging of hydrogen peroxide. Analytical Chemistry. 2016;**88**(9):4833-4840. DOI: 10.1021/acs.analchem.6b00521

[72] Wang L, Li W, Li M, Su Q, Li Z, Pan D, et al. Ultrastable amine, sulfo cofunctionalized graphene quantum dots with high two-photon fluorescence for cellular imaging. ACS Sustainable Chemistry & Engineering. 2018;**6**(4):4711-4716. DOI: 10.1021/acssuschemeng.7b03797

[73] Sapkota B, Benabbas A, Lin H-YG, Liang W, Champion P, Wanunu M. Peptide-decorated tunable-fluorescence graphene quantum dots. ACS Applied Materials & Interfaces. 2017;**9**(11):9378-9387. DOI: 10.1021/acsami.6b16364

[74] Chen T-T, Tian X, Liu C-L, Ge J, Chu X, Li Y. Fluorescence activation imaging of cytochrome c released from mitochondria using aptameric nanosensor. Journal of the American Chemical Society. 2015;**137**(2):982-989. DOI: 10.1021/ja511988w

[75] Wang L, Zheng J, Yang S, Wu C, Liu C, Xiao Y, et al. Two-photon sensing and imaging of endogenous biological cyanide in plant tissues using graphene quantum dot/gold nanoparticle conjugate. ACS Applied Materials & Interfaces. 2015;**7**(34):19509-19515. DOI: 10.1021/acsami.5b06352

[76] Hong M, Xu L, Xue Q, Li L, Tang B. Fluorescence imaging of intracellular telomerase activity using enzyme-free signal amplification. Analytical Chemistry. 2016;**88**(24):12177-12182. DOI: 10.1021/acs.analchem.6b03108

[77] Liu Q, Guo B, Rao Z, Zhang B, Gong JR. Strong two-photon-induced fluorescence from photostable, biocompatible nitrogen-doped graphene quantum dots for cellular and deep-tissue imaging. Nano Letters. 2013;**13**(6):2436-2441. DOI: 10.1021/nl400368v

[78] Campbell E, Hasan MT, Gonzalez Rodriguez R, Akkaraju GR, Naumov AV. Doped graphene quantum dots for intracellular multicolor imaging and cancer detection. ACS Biomaterials Science & Engineering. 2019;**5**(9):4671-4682. DOI: 10.1021/acsbiomaterials.9b00603

[79] Wu C, Li P, Fan N, Han J, Zhang W, Zhang W, et al. A dual-targeting functionalized graphene film for rapid and highly sensitive fluorescence imaging detection of hepatocellular carcinoma circulating tumor cells. ACS Applied Materials & Interfaces. 2019;**11**(48):44999-45006. DOI: 10.1021/acsami.9b18410

[80] Wang H-B, Ou L-J, Huang K-J, Wen X-G, Wang L-L, Liu Y-M. A sensitive biosensing strategy for DNA detection based on graphene oxide and T7 exonuclease assisted target recycling amplification. Canadian Journal of Chemistry. 2013;**91**(12):1266-1271. DOI: 10.1139/cjc-2013-0285

[81] Sharma MD, Rayalu SS, Kolev SD, Krupadam RJ. Graphene/fluorescein dye-based sensor for detecting As(III) in drinking water. Scientific Reports. 2021;**11**(1):17321. DOI: 10.1038/s41598-021-96968-3

[82] Youn H, Lee K, Her J, Jeon J, Mok J, So J-I, et al. Aptasensor for multiplex detection of antibiotics based on FRET

strategy combined with aptamer/graphene oxide complex. Scientific Reports. 2019;**9**(1):7659. DOI: 10.1038/s41598-019-44051-3

[83] Zheng X, Zhai R, Zhang Z, Zhang B, Liu J, Razaq A, et al. Graphene-oxide-based fluoro- and chromo-genic materials and their applications. Molecules. 2022;**27**(6):2018. DOI: 10.3390/molecules27062018

[84] Nurunnabi M, Khatun Z, Reeck GR, Lee DY, Lee Y-K. Photoluminescent graphene nanoparticles for cancer phototherapy and imaging. ACS Applied Materials & Interfaces. 2014;**6**(15):12413-12421. DOI: 10.1021/am504071z

[85] Narasimhan AK, Lakshmi BS, Santra TS, Rao MSR, Krishnamurthi G. Oxygenated graphene quantum dots (GQDs) synthesized using laser ablation for long-term real-time tracking and imaging. RSC. Advances. 2017;**7**(85):53822-53829. DOI: 10.1039/c7ra10702a

[86] Liu J-W, Wang Y-M, Zhang C-H, Duan L-Y, Li Z, Yu R-Q, et al. Tumor-targeted graphitic carbon nitride nanoassembly for activatable two-photon fluorescence imaging. Analytical Chemistry. 2018;**90**(7):4649-4656. DOI: 10.1021/acs.analchem.7b05192

[87] Nahain A-A, Lee J-E, In I, Lee H, Lee KD, Jeong JH, et al. Target delivery and cell imaging using hyaluronic acid-functionalized graphene quantum dots. Molecular Pharmaceutics. 2013;**10**(10):3736-3744. DOI: 10.1021/mp400219u

[88] Coney LR, Tomassetti A, Carayannopoulos L, Frasca V, Kamen BA, Colnaghi MI, et al. Cloning of a tumor-associated antigen: MOv18 and MOv19 antibodies recognize a folate-binding protein. Cancer Research. 1991;**51**(22):6125-6132

[89] Goreham RV, Schroeder KL, Holmes A, Bradley SJ, Nann T. Demonstration of the lack of cytotoxicity of unmodified and folic acid modified graphene oxide quantum dots, and their application to fluorescence lifetime imaging of HaCaT cells. Mikrochimica Acta. 2018;**185**(2):128. DOI: 10.1007/s00604-018-2679-8

[90] Kumawat MK, Thakur M, Gurung RB, Srivastava R. Graphene quantum dots for cell proliferation, nucleus imaging, and photoluminescent sensing applications. Scientific Reports. 2017;**7**(1):15858. DOI: 10.1038/s41598-017-16025-w

[91] Fan Z, Zhou S, Garcia C, Fan L, Zhou J. pH-responsive fluorescent graphene quantum dots for fluorescence-guided cancer surgery and diagnosis. Nanoscale. 2017;**9**(15):4928-4933. DOI: 10.1039/c7nr00888k

[92] Lu H, Li W, Dong H, Wei M. Graphene quantum dots for optical bioimaging. Small. 2019;**15**(36):e1902136. DOI: 10.1002/smll.201902136

[93] Zhang M, Cao Y, Chong Y, Ma Y, Zhang H, Deng Z, et al. Graphene oxide based theranostic platform for T1-weighted magnetic resonance imaging and drug delivery. ACS Applied Materials & Interfaces. 2013;**5**(24):13325-13332. DOI: 10.1021/am404292e

[94] Yang Y, Chen S, Li H, Yuan Y, Zhang Z, Xie J, et al. Engineered paramagnetic graphene quantum dots with enhanced relaxivity for tumor imaging. Nano Letters. 2019;**19**(1):441-448. DOI: 10.1021/acs.nanolett.8b04252

[95] Mieog JSD, Achterberg FB, Zlitni A, Hutteman M, Burggraaf J,

Swijnenburg R-J, et al. Fundamentals and developments in fluorescence-guided cancer surgery. Nature Reviews. Clinical Oncology. 2022;**19**(1):9-22. DOI: 10.1038/s41571-021-00548-3

[96] Lee Y-J, van den Berg NS, Orosco RK, Rosenthal EL, Sorger JM. A narrative review of fluorescence imaging in robotic-assisted surgery. Laparoscopic Surgery. 2021;**5**:31-31. DOI: 10.21037/ls-20-98

[97] Egloff-Juras C, Bezdetnaya L, Dolivet G, Lassalle H-P. NIR fluorescence-guided tumor surgery: New strategies for the use of indocyanine green. International Journal of Nanomedicine. 2019;**14**:7823-7838. DOI: 10.2147/IJN.S207486

[98] Chen J, Liu C, Zeng G, You Y, Wang H, Gong X, et al. Indocyanine green loaded reduced graphene oxide for in vivo photoacoustic/fluorescence dual-modality tumor imaging. Nanoscale Research Letters. 2016;**11**(1):85. DOI: 10.1186/s11671-016-1288-x

[99] Cui F, Ji J, Sun J, Wang J, Wang H, Zhang Y, et al. A novel magnetic fluorescent biosensor based on graphene quantum dots for rapid, efficient, and sensitive separation and detection of circulating tumor cells. Analytical and Bioanalytical Chemistry. 2019;**411**(5):985-995. DOI: 10.1007/s00216-018-1501-0

[100] Han Q, Pang J, Li Y, Sun B, Ibarlucea B, Liu X, et al. Graphene biodevices for early disease diagnosis based on biomarker detection. ACS Sensors. 2021;**6**(11):3841-3881. DOI: 10.1021/acssensors.1c01172

[101] Fahmy HM, Abu Serea ES, Salah-Eldin RE, Al-Hafiry SA, Ali MK, Shalan AE, et al. Recent progress in graphene- and related carbon-nanomaterial-based electrochemical biosensors for early disease detection. ACS Biomaterials Science & Engineering. 2022;**8**(3):964-1000. DOI: 10.1021/acsbiomaterials.1c00710

Chapter 4

Optical Chemosensors: Principles, Chemistry, Strategies, and Applications

Mohamed Yahya, Müjgan Yaman and Zeynel Seferoğlu

Abstract

Chemosensors for anions and cations detections have been extensively used in several disciplines, including pharmacology, environmental science, biology, and chemistry. This field which is a division of supramolecular chemistry has been known for more than 150 years. It deals with chemosensors that recognize and detect anions and cations via optical or electrochemical signals. Today, a sustainable variety of chemosensors are established to detect both anions and cations. Additionally, chemosensors can be used to construct a sensory device and extract, and separate anions and cations. Chemosensors can detect toxic anions such as fluoride and cyanide as well as cations like mercury. Thus, chemosensors have become an attractive area of supramolecular chemistry. This chapter focuses on both colorimetric and fluorometric optical chemosensors and their application for anions and cations detections.

Keywords: chemosensors, colorimetric, fluorescence, anions and cations detection, optical chemosensors

1. Introduction

Czarnick defines the chemosensor as a "molecule of abiotic origin that signals the presence of matter or energy [1]." Furthermore, chemosensors are molecule receptors that can sense and precisely interact with an analyte and generate a response as a detectable signal. The signal can be optical or electrical. It consists of a detection group and a signaling moiety. The detection group is responsible for selectivity and binding efficiency. While converting information into a detectable signal is due to the signaling moiety.

Recently, considerable studies focused on chemosensors because of their susceptible photophysical properties to the environment. The change in the optical signals can provide data on the chemical parameters such as pH and analytes concentration [2]. Chemosensors are extensively investigated to detect heavy metal ions [3, 4]. Optical Chemosensors are classified into colorimetric and Fluorometric chemosensors. The colorimetric receptors display the advantage of naked-eye detection of color changes, making the process simple.

On the other hand, the Fluorometric chemosensors are more chosen for the ratiometric response because of the ratio between the intensity of two emissions

IntechOpen

which allows for correcting the analytic concentration of the sensor and the effects of environment like temperature and polarity [5]. Today the demand is high for chemosensors that are selective towards specific harmful ions. The toxicity of certain anions and cations for humans as well as animals has motivated researchers to design chromophores that are selective to a specific anion or cation [6–8]. Several international agencies prohibit numerous heavy metal ions because of their toxicity and non-biodegradability, resulting in their accumulation in the environment [9, 10]. Nowadays, chemists and scientists are attempting to design chemosensors that can be used for environmental and industrial sample analysis.

Fluorometric chemosensors have attracted increasing intention for detecting selective anion or cation due to:

- High Sensitivity (single-molecule detection is probable)
- High selectivity
- Short response time
- High spatial and temporal resolution
- Low cost and easily performed instrumentations

In this contribution, both colorimetric and fluorometric chemosensors will be discussed, focusing on fluorometric receptors. The most crucial applications for Fluorimetric chemosensors for detecting Cyanide and Fluoride are also briefly given.

2. Optical chemosensors

Optical sensors are investigation techniques that detect light intensity. They modify the receptor's photophysical properties upon analyte (guest) binding to the receptor. Depending on the sensor type, these modifications are characterized in UV–visible and fluorescence spectroscopy instruments. The chemosensors can be classified into colorimetric chemosensors, fluorometric chemosensors, or fluorescent chemosensors.

2.1 Colorimetric chemosensors

A colorimetric chemosensor is defined as the color change that occurs after the receptor's binding with a specific analyte [11]. After binding the receptor and the analyte, the chemosensor's signaling unit shows a color change [12]. Colorimetric chemosensors have attracted significant interest due to the possibility of obtaining qualitative and quantitative data via the naked eye [13] without referring to any complex techniques [14].

2.2 Fluorometric chemosensors

2.2.1 General principle: design of chemosensor

One of the most valuable response methods for optical readout is fluorescence. The fluorescence chemosensors have attracted considerable interest due to their sensitivity, selectivity, quick response time, on-site and real-time detection, straightforward

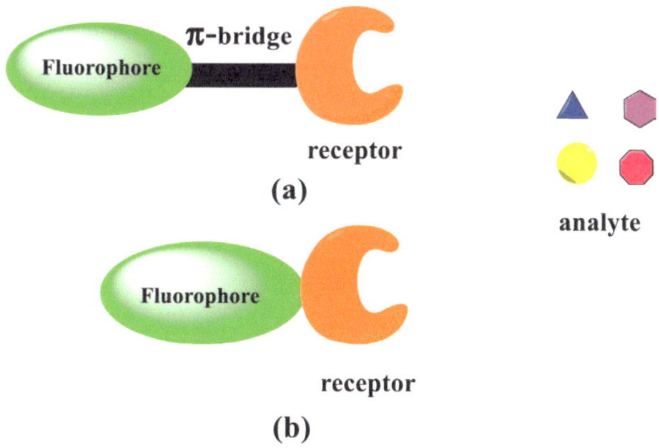

Figure 1.
Fluorescence sensing (a) A spaced model (b) An integrated model.

performance, flexibility, and present low molar estimate of the analyte. In the case of fluorometric chemosensors, binding the analyte to the receptor leads to a change in fluorescence behavior [15]. In 1867, Goppelsroder reported the first fluorescent chemosensor to detect aluminum ions via the formation of a morin chelate which is strongly fluorescent.

The fluorescent chemosensor allows the detection at the picomolar scale; however, the colorimetric sensors can only detect concentration at micromolar levels. The fluorometric chemosensor's advantages are mainly due to the proportionality of the emitted fluorescence to the analyte concentration. In contrast, in absorbance measurements, the concentration of the analyte and the absorbance are proportional, which is associated with the ratio between intensities measured before and after the beam passes through the sample.

The essential parts of the fluorescent chemosensors are the recognition and signaling moieties, as given in **Figure 1**. Accordingly, a fluorescent chemosensor is developed by connecting a receptor (ionophore) to a fluorophore responsible for converting the recognition into the photophysical signal (like spectra, fluorescence quantum yield, and lifetime).

The photophysical properties can alter by binding a particular analyte to the receptor, leading to a fluorescence signal, with either an enhancement or quenching of fluorescence. The binding of the analyte to the receptor leads to an enhanced fluorescence intensity called turn-on chemosensor [16, 17], whereas the analyte bounding to the fluorophore results in a decrease of fluorescent intensity quenching called turn-off chemosensor [18, 19].

Depending on the connection of fluorophore and receptor, two models of fluorescent chemosensors can be defined: spaced (**Figure 1(a)**) and integrated model (**Figure 1(b)**). The spaced model offers a design where the fluorophore is connected to the receptor via spacer and signaling moieties that prevent conjugations. In the integrated model, the fluorophore and receptor are conjugatively connected to each other; in this model, the receptor is a part of the π-electron system of the fluorophore.

2.2.2 Photo-induced electron transfer

The fluorescence quenching is due to a photo-induced electron transfer (PET); the normal state returns when a non-luminescent process follows the PET process [10].

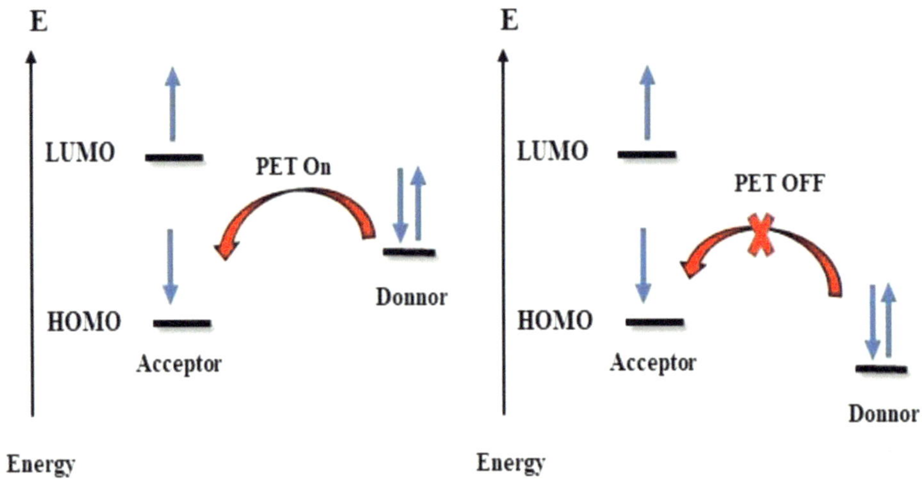

Figure 2.
Molecular orbital diagram of the photo-induced electron transfer process (PET).

The PET process could explain the "on–off" switching of fluorescent chemosensors based on the molecular orbital theory. The basic design of the florescent chemosensors based on PET is a fluorophore- spacer- receptor. This model has a spacer separating the fluorophore from the receptor, consequently electronically disconnecting the π-electron systems from the fluorophore and receptor. The PET process is divided into two types: reductive PET and oxidative PET, depending on the electron acceptor or donor links to the fluorophore and receptor [11]. However, in this study, only the reductive PET will be investigated. The design of the fluorescent chemosensors is given in **Figure 2**. The reductive PET is defined as the system where the fluorophore is reduced, whereas the receptor is oxidized. In this type of PET process, the fluorophore provides an electron acceptor, and the receptor serves as an electron donor. The electron photo-excitation from HOMO to LUMO occurs through the reductive PET process. Consequently, the fluorophore and the analyte react and cause the occurrence or elimination HOMO and LUMO "adjacent" orbital, resulting in the fluorescence quenching or enhancement, respectively (**Figure 3**).

2.2.3 Intra and intermolecular charge transfer

Valeur et al. were the first to introduce the intramolecular charge transfer (ICT) process for cation sensing [12]. ICT process is defined as an excited state where the conjugation of an electron-donating unit (such as $-NH_2$, $-NMe_2$, $-OCH_3$) to an electron-accepting (like >C=O, –CN) unit in one molecule is shown to rise a "push-pull" π- electron system [13], which have been widely applied for cation sensing. A blue shift in the absorption spectrum is observed when the electron donor interacts with the analyte, decreasing the electron-donating character. However, the analyte reacts with the electron acceptor part leading to a red shift due to a developed ICT (**Figure 4**). Furthermore, alterations in the fluorescence lifetimes and quantum yields are also detected. Thus far, many fluorescent imaging molecules are obtained from the ICT process by modifying either the electron donor, electron acceptor capacity, or π-conjugation degree of the fluorophores to react with the target analyte.

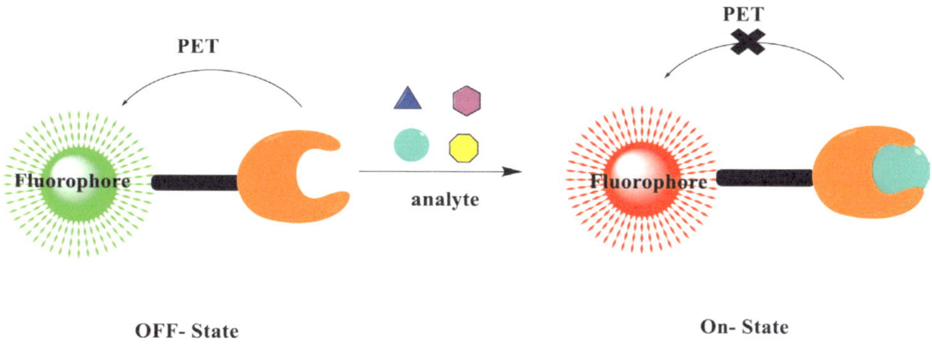

Figure 3.
Schematic representation of PET chemosensor in the sensing process.

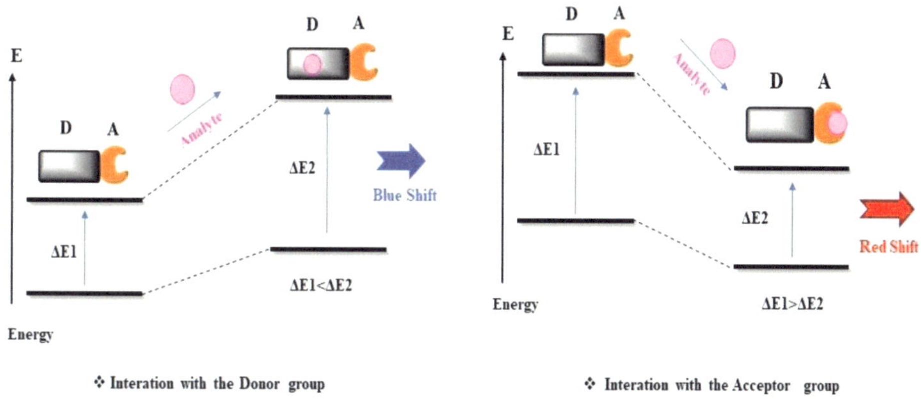

Figure 4.
Schematic representation of ICT chemosensor in the sensing process.

2.2.4 Excited-state intramolecular proton transfer

Excited-state intramolecular proton transfer (ESIPT) has been implemented to design fluorescent chemosensors due to their distinctive and exceptional spectral sensitivity to the environmental medium. ESIPT process is based on a proton transfer from a proton donor (hydroxyl or amino unit) to an acceptor unit (carbonyl oxygen or imine nitrogen) atom in the excited state of a fluorophore which is facilitated by an intramolecular hydrogen bond [20]. A graphical interpretation of the ESIPT process is illustrated in **Figure 5**.

Due to the change between the enol and keto form in the system based on ESIPT, the photochemical reactivity of the excited molecules is drastically reduced, leading to improved photostability. Additionally, a significant Stokes shift is observed. Hence, the ESIPT process fits the design of fluorescent chemosensors that necessitates spectral shift for selective detection.

2.2.5 Fluorescence resonance energy transfer

Fluorescence Resonance Energy Transfer (FRET) is another type of fluorescence modulation. The FRET process is a transfer of energy between a pair of fluorophores

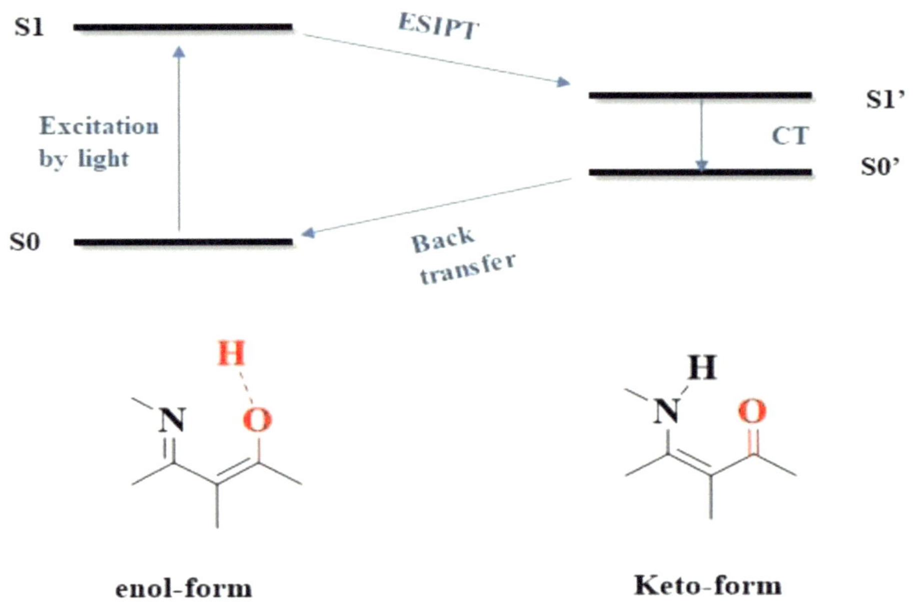

Figure 5.
Schematic representation of excited-state intramolecular proton transfer system.

that operate as energy donors and acceptors, respectively [21, 22]. The FRET depends on the interaction distance between the electronically excited states of two chromophores that the excitation energy is non-radiatively relocated from a donor to an acceptor by non-radiative dipole–dipole coupling (**Figure 6**). The vibrational transitions in the donor and the acceptor are approximately equivalent. If the FRET is operative in the molecules, they are typically applied to improve the stokes shift artificially. For sensing applications, the emission of the donor at relatively short wavelengths leads to triggering the acceptor emission at longer wavelengths with a ratio of the fluorescence intensities of the donor and acceptor emissions regulated by the target analytes. Elevated FRET efficacy is obtained when the extensive spectral overlaps with the donor emission and the acceptor absorption spectrum [23]. Several FRET-based fluorescent chemosensors were developed following the groundbreaking endeavors for the ratiometric detection of metal ions like Hg^{2+} [24], Zn^{2+} [25].

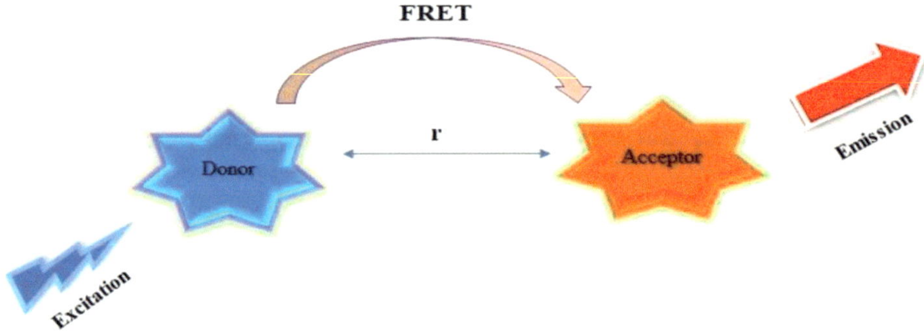

Figure 6.
Schematic representation of FRET chemosensor in the sensing process.

3. Applications of fluorescent chemosensors

Cyanide (CN^-) and Fluoride (F^-) play an essential role in life and are deemed one of very poisonous anions and harmful to human health. They are used in industrial activities like acrylic fiber manufacturing, metallurgy, electroplating, fibers synthesis, and gold extraction. Absorption by the lungs, exposure to the skin, food, and contaminated drinking water are all exposition methods to CN^- and F^- intoxication. **Table 1** summarizes the World Health Organization (WHO) and European (EU) guidelines for anions in drinking water. The WHO set the provisional maximum tolerable daily intake limits to 0.07 mg/L and 1.5 mg/L for CN^- and F^-, respectively, for drinking water. It is consequently of paramount importance to be able to establish their concentration in water at such low levels precisely. The detection of anions is generally effectuated via analytic techniques such as Ionic chromatography. However, these techniques are complex and expensive, where the need to develop new techniques for anions detection that are both accurate and low cost. Recently, fluorescent chemosensors have attracted considerable attention as a promising alternative for anions sensing.

The fluorescent chemosensors have been used in numerous fields, such as biology, physiology, pharmacology, and environmental sciences. Fluorescent chemosensors to detect environmentally critical contaminants such as anions (CN^-, F^-) or cations (Hg^+, Pb^{2+} ...) have been widely investigated.

Keleş et al. reported the synthesis of 3,5-dinitro-(*N*-phenyl)benzamide (DNBA) and it is used as a simple colorimetric and fluorimetric chemosensor for selective determination of CN- in organic (DMSO and ACN) and aqueous solutions (DMSO/H_2O: 8:2, v/v) [26]. The colorimetric and fluorimetric color intensity changed after adding CN due to the interaction between CN- and DNBA. The UV/Vis and fluorescence spectrometry defined the interaction mechanisms between DNBA and CN-. Furthermore, spectroscopic results showed that CN- interacts with three different mechanisms; deprotonation, nucleophilic aromatic substitution, and formation of benzisoxazole ring. **Figure 7** displays the detection mechanism after adding CN- in a DMSO/H_2O medium.

Zheng et al. studied the use of naphthalimide-based (NIMS) as a fluorescent sensor for F^- [27]. The authors reported that F- detection occurs via a desilylation reaction (**Figure 8**), resulting in a colorimetric/fluorometric spectral response with a broad absorption around 229 nm and a change in color from yellow to blue. Furthermore, it was found that NIMS have high selectivity, rapid response, and sound sensitivity for F^-. NIMS is a promising sensor for F^- detection and quantification in toothpaste.

Element	Symbol	WHO standards (mg/L)	EU standards (mg/L)
Cyanide	CN^-	0.07	0.05
Fluoride	F^-	1.5	1.5
Chloride	Cl^-	250	250
Hydrogen sulfate	HSO_4^-	No guideline	No guideline
Bromine	Br^-	Not mentioned	0.01
Nitrate	NO_3^-	50	50

Table 1.
WHO and EU drinking water standards for anions.

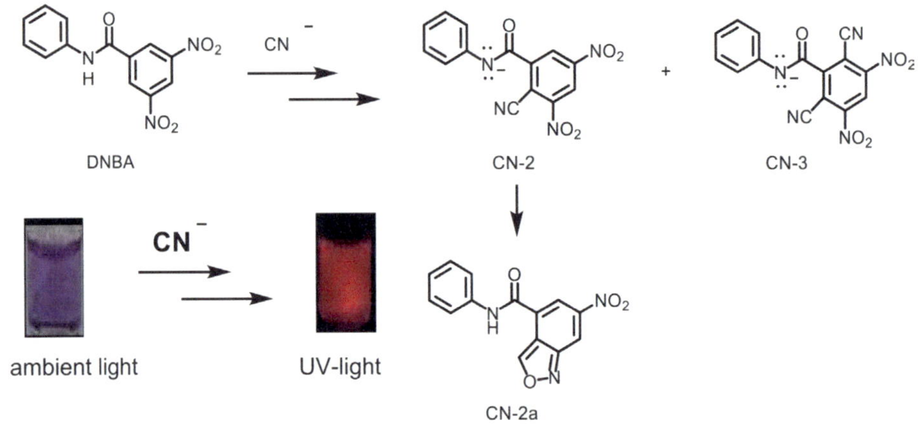

Figure 7.
CN^{-} detection mechanism reported by Keleş et al. [26].

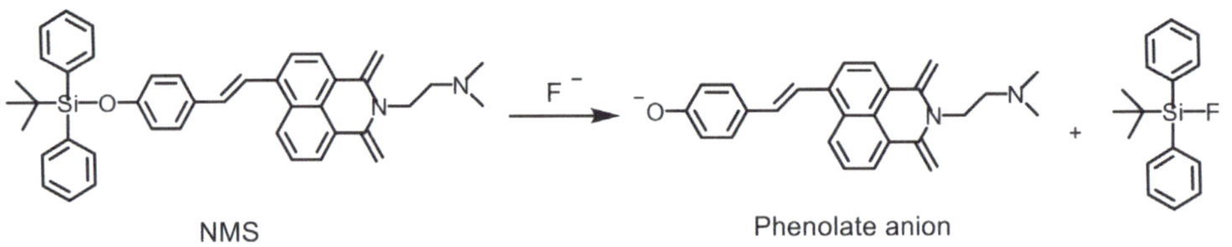

Figure 8.
Mechanism of F^{-} detection using NIMS sensor.

Sourav Bej et al. reported the synthesis of an oxene-based chemosensor (HyMa) via Knoevenagel condensation to detect multi-analytes such as HSO_4^{-}, CN^{-}, and F^{-} [28]. The authors reported a detection limit of around 38 ppb, 18 ppb, and 94 ppb for HSO_4^{-}, CN^{-}, and F^{-} respectively. Furthermore, 3-(4,5-dimethylthiazol-2-yl)-2,5-diphenyl tetrazolium bromide (MTT) displayed a low cytotoxicity and membrane permeability, making it an attractive material for bio-imaging. The mechanism of CN^{-} detection is given in **Figure 9**.

Deng et al. reported another chemosensor for the detection of CN^{-}. TCNT fluorescent displaying an aggregation-induced emission was synthesized. A red fluorescent emission (596 nm) and significant stokes shift (148 nm) were observed, leading to an increased sensitivity to CN^{-} (detection limit = 0.38 μmol $L-^{1}$) [29]. The detection was due to a nucleophilic attack of CN^{-} on the vinyl group. Furthermore, the authors proved that TCNT could be used for bioimaging in live organisms.

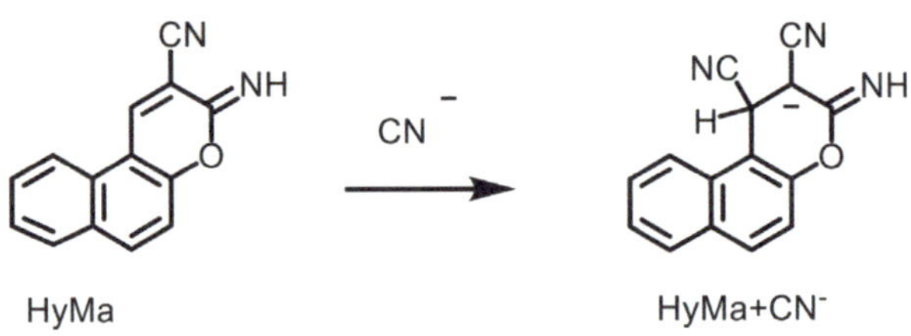

Figure 9.
The detection mechanism of CN^{-} by HyMa.

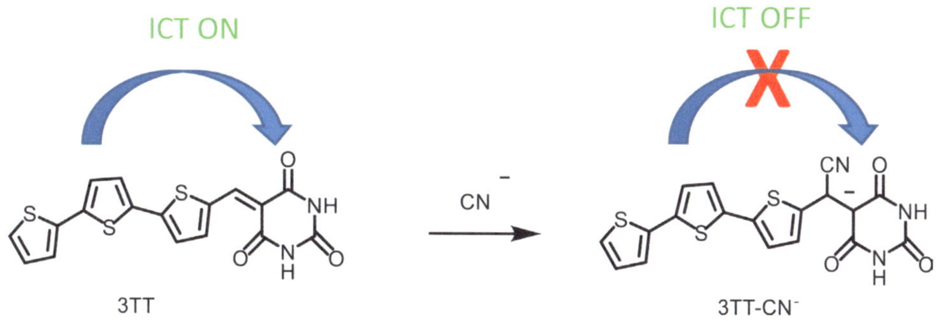

Figure 10.
The detection mechanism of CN^- by 3TT.

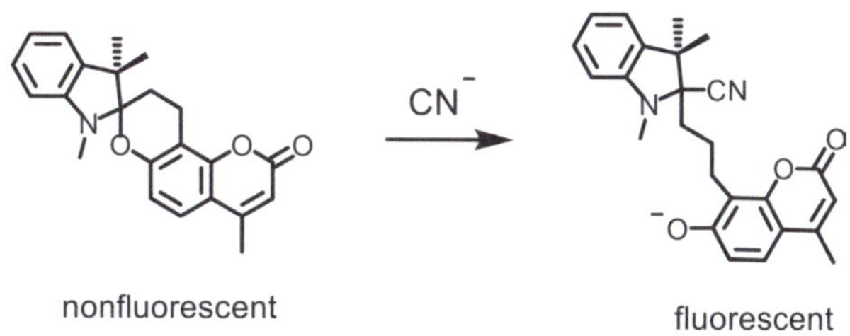

Figure 11.
The detection mechanism of CN^-.

A donor-π-acceptor structure was synthesized by Sun et al. for the CN^- detection (3TT). The detection of CN^- occurs via nucleophilic addition to the β-conjugated position of the barbituric acid of 3TT (**Figure 10**). The authors reported a naked-eye detection with high anti-interference ability and quick response. The detection limit was found to be 39.9 nM [24]. Furthermore, 3TT exhibited exceptional performance in the solid-state, and the 3TT-based test filter paper strips were applied to quickly and effectively detect CN^- in water with naked eyes. Likewise, the 3TT was effectively applied to rapidly detect CN^- in actual water samples, sour seeds, and food samples, proving its good ability for practical applications in our daily life.

Coumarin-spiropyran dyad having a hydrogen pyran moiety [2] was synthesized by [25]. The structure has an off–on type fluorescent receptor for detecting CN^-. Because of the delocalization of π-electrons over the molecule, the receptor has nearly no fluorescence with a quantum yield <0.01. The nucleophilic addition of CN- to the spirocarbon of the molecule leads to a rapid opening of the spirocycle (**Figure 11**). Hence the localization of π-electrons on the coumarin moiety and a strong light-blue fluorescence occurs at 459 nm, and a high quantum yield (0.52) is obtained. The detection limit was low and equal to 1.0 μM.

4. Conclusions

Since Goppelsröder stated the first fluorescent chemosensors selective for Al^{3+}, the chemosensors field has progressed extensively. Notably, the last 50 years witnessed massive development in the field of fluorometric chemosensors. Hence, numerous chemosensors have been designed by researchers over the last years. Chemosensors present a promising technology for detecting toxic anions and cations in an aqueous medium. This contribution gave a summary of the chemosensors' chemistry and their applications. Finally, some examples of cyanide and fluoride, among other examples, detection were given.

Acknowledgements

This research was supported by Gazi University. This work was a part of Dr. Mohamed Yahya thesis of Reference No: 10451335.

Conflict of interest

The authors declare no conflict of interest.

Author details

Mohamed Yahya[1,2]*, Müjgan Yaman[3] and Zeynel Seferoğlu[1,4]

1 Faculty of Science, Department of Chemistry, Gazi University, Ankara, Turkey

2 Department of Chemistry, University of Nevada, Reno, NV, USA

3 Faculty of Science, Department of Chemistry, Eskisehir Osmangazi University, Eskisehir, Turkey

4 Technological Dyes and Materials Application and Research Center (TEBAM), Gazi University, Ankara, Turkey

*Address all correspondence to: myahya@umr.edu

References

[1] Czarnik AW. Supramolecular chemistry, fluorescence, and sensing. In: Fluorescent Chemosensors for Ion and Molecule Recognition UTC; 1993. p. 1-9. [cited June 9, 2022]. DOI: 10.1021/bk-1993-0538.ch001

[2] Liu S, Wang YM, Han J. Fluorescent chemosensors for copper(II) ion: Structure, mechanism and application. Journal of Photochemistry and Photobiology C Photochemistry Reviews. 2017;**32**:78-103

[3] Patil NS, Dhake RB, Ahamed MI, Fegade U. A mini review on organic chemosensors for cation recognition (2013-2019). Journal of Fluorescence. 2020. cited 10 May 2022;**30**(6): 1295-1330. Available from: https://link.springer.com/article/10.1007/s10895-020-02554-7

[4] Kowser Z, Rayhan U, Akther T, Redshaw C, Yamato T. A brief review on novel pyrene based fluorometric and colorimetric chemosensors for the detection of Cu2+ Vol. 5, Materials Chemistry Frontiers. Royal Society of Chemistry; 2021 [cited June 10, 2021]. p. 2173-2200. Available from: https://pubs.rsc.org/en/content/articlehtml/2021/qm/d0qm01008a

[5] Kollur SP, Shivamallu C, Prasad SK, Veerapur R, Patil SS, Cull CA, et al. Recent Advances on the development of chemosensors for the detection of mercury toxicity: A review. 2021;**8**(10):192. [cited May 10, 2022] Available from: https://www.mdpi.com/2297-8739/8/10/192/htm

[6] Li MM, Huang SY, Ye H, Ge F, Miao JY, Zhao BX. A new pyrazoline-based fluorescent probe for Cu2+ in live cells. Journal of Fluorescence. 2013 [cited 10 May 2022;**23**(4):799-806. Available from: https://link.springer.com/article/10.1007/s10895-013-1203-0

[7] Ma S, Wang Y, She M, Wang S, Yang Z, Liu P, et al. Design strategies and progress on xanthene-based fluorescent probe for metal ions. Reviews in Analytical Chemistry 2017. cited 10, May 2022;**36**(2):20160024. Available from: https://www.degruyter.com/document/doi/10.1515/revac-2016-0024/html?lang=en

[8] Kaur N, Kaur G, Fegade UA, Singh A, Sahoo SK, Kuwar AS, et al. Anion sensing with chemosensors having multiple NH recognition units. TrAC Trends in Analytical Chemistry. 2017;**95**:86-109

[9] DIRECTIVE 2002/95/EC OF THE EUROPEAN PARLIAMENT AND OF THE COUNCIL of 27 January 2003 on the restriction of the use of certain hazardous substances in electrical and electronic equipment. Official Journal of the European Union.

[10] WHO (2004) Guidelines for Drinking Water Quality. 3rd ed. Vol. 1. Geneva: World Health Organization; 2004. p.188. [cited May 10,2022]. Available from: https://books.google.com.tr/books?hl=en&lr=&id=SJ76COTm-nQC&oi=fnd&pg=PR15&dq=World+Health+Organization+(2004)+Guidelines+for+drinking-water+quality+(3rd+ed)+vol+1,+Geneva,+188&ots=V9tWpfPa0d&sig=fltvRqdL4WJrZH_jkE0SnoZxrn0&redir_esc=y#v=onepage&q&f=false

[11] Wang M, Leung KH, Lin S, Chan DSH, Kwong DWJ, Leung CH, et al. A colorimetric chemosensor for CU2+ ion detection based on an iridium(III) complex. Scientific Reports. 2014;**4**(1):1-7

[12] Kaur P, Kaur S, Singh K. Colorimetric detection of cyanide in water using a highly selective Cu2+ chemosensor. Inorganic Chemistry Communications. 2009;**12**(10):978-981

[13] Dong X, Wang R, Liu G, Liu P, Pu S. A novel sensitive sensor for Cu2+ and multi-switch based on a diarylethene with a 2-(2′-hydroxyphenyl) benzothiazole unit. Tetrahedron. 2016;**72**(22):2935-2942

[14] Park GJ, Lee JJ, You GR, Nguyen L, Noh I, Kim C. A dual chemosensor for Zn2+ and Co2+ in aqueous media and living cells: Experimental and theoretical studies. Sensors and Actuators B: Chemical. 2016;**223**:509-519

[15] Wang JH, Liu YM, Dong ZM, Bin CJ, Wang H, Wang Y, et al. New colorimetric and fluorometric chemosensor for selective Hg2+ sensing in a near-perfect aqueous solution and bio-imaging. Journal of Hazardous Materials. 2020;**382**:121056

[16] Zeng S, Li SJ, Liu TT, Sun XJ, Xing ZY. A significant fluorescent "turn-on" chemosensor for Al3+ detection and application in real sample, logic gate and bioimaging. Inorganica Chimica Acta. 2019;**495**:118962

[17] Gupta A, Kumar N. A review of mechanisms for fluorescent "turn-on" probes to detect Al3+ ions. RSC Advances. Royal Society of Chemistry. 2016;**6**:106413-106434

[18] Riis-Johannessen T, Schenk K, Severin K. Turn-off-and-on: Chemosensing ensembles for sensing chloride in water by fluorescence spectroscopy. Inorganic Chemistry. 2010;**49**(20):9546-9553

[19] Wu D, Sedgwick AC, Gunnlaugsson T, Akkaya EU, Yoon J, James TD. Fluorescent chemosensors: The past, present and future. Chemical Society Reviews. Royal Society of Chemistry. 2017;**46**:7105-7123

[20] Yang Y, Zhao Q, Feng W, Li F. Luminescent chemodosimeters for bioimaging. Chemical Reviews. American Chemical Society. 2013;**113**:192-270

[21] Fan J, Hu M, Zhan P, Peng X. Energy transfer cassettes based on organic fluorophores: Construction and applications in ratiometric sensing. Chemical Society Reviews. 2013;**42**(1):29-43

[22] Kim JS, Quang DT. Calixarene-derived fluorescent probes. Chemical Reviews. American Chemical Society. 2007;**107**:3780-3799

[23] Feng Y, Cheng J, Zhou L, Zhou X, Xiang H. Ratiometric optical oxygen sensing: A review in respect of material design. Analyst. Royal Society of Chemistry. 2012;**137**:4885-4901

[24] Sun T, Niu Q, Li Y, Li T, Hu T, Wang E, et al. A novel oligothiophene-based colorimetric and fluorescent "turn on" sensor for highly selective and sensitive detection of cyanide in aqueous media and its practical applications in water and food samples. Sensors and Actuators B: Chemical. 2018;**258**:64-71

[25] Shiraishi Y, Nakamura M, Hayashi N, Hirai T. Coumarin-spiropyran dyad with a hydrogenated pyran moiety for rapid, selective, and sensitive fluorometric detection of cyanide anion. Analytical Chemistry. 2016 [cited June 9, 2022]. 88(13):6805-6811. DOI 10.1021/acs.analchem.6b01279

[26] Keleş E, Aydıner B, Nural Y, Seferoğlu N, Şahin E, Seferoğlu Z. A new mechanism for selective recognition of

cyanide in organic and aqueous solution. European Journal of Organic Chemistry. 2020;(30):4681-4692. [cited June 6, 2022]. DOI: 10.1002/ejoc.202000342

[27] Zheng X, Zhu W, Ai H, Huang Y, Lu Z. A rapid response colorimetric and ratiometric fluorescent sensor for detecting fluoride ion, and its application in real sample analysis. Tetrahedron Letters. 2016;**57**(52):5846-5849

[28] Bej S, Das R, Mondal A, Saha R, Sarkar K, Banerjee P. Knoevenagel condensation triggered synthesis of dual-channel oxene based chemosensor: Discriminative spectrophotometric recognition of F−, CN− and HSO4− with breast cancer cell imaging, real sample analysis and molecular keypad lock applications. Spectrochimica Acta, Part A: Molecular and Biomolecular Spectroscopy. 2022;**273**:120989

[29] Deng K, Wang L, Xia Q, Liu R, Qu J. A turn-on fluorescent chemosensor based on aggregation-induced emission for cyanide detection and its bioimaging applications. Sensors and Actuators B: Chemical. 2019;**296**:126645

Section 3

Novel Fluorescence Imaging Techniques and Methodologies

Chapter 5

Lifetime Determination Algorithms for Time-Domain Fluorescence Lifetime Imaging: A Review

Yahui Li, Lixin Liu, Dong Xiao, Hang Li, Natakorn Sapermsap, Jinshou Tian, Yu Chen and David Day-Uei Li

Abstract

Fluorescence lifetime imaging (FLIM) is powerful for monitoring cellular microenvironments, protein conformational changes, and protein interactions. It can facilitate metabolism research, drug screening, DNA sequencing, and cancer diagnosis. Lifetime determination algorithms (LDAs) adopted in FLIM analysis can influence biological interpretations and clinical diagnoses. Herein, we discuss the commonly used and advanced time-domain LDAs classified in fitting and non-fitting categories. The concept and explicit mathematical forms of LDAs are reviewed. The output lifetime parameter types are discussed, including lifetime components, average lifetimes, and graphic representation. We compare their performances, identify trends, and provide suggestions for end users in terms of multi-exponential decay unmixing ability, lifetime estimation precision, and processing speed.

Keywords: fluorescence lifetime imaging, lifetime determination algorithm, fitting method, non-fitting method, deep learning

1. Introduction

Fluorescence lifetime imaging (FLIM) is a vital and versatile technique for assessing molecular microenvironments of fluorophores in living cells, such as pH, O_2, viscosity, temperature, or ion concentrations [1, 2]. FLIM can be a powerful "quantum ruler" to measure subnanometer protein conformational changes and interactions by quantifying the occurrence of Förster Resonance Energy Transfer (FRET) [3–7]. FLIM has been used in diverse disciplines, including biology, chemistry and biophysics [8–12]; however, it is an indirect imaging technique that needs sophisticated data analysis to deliver meaningful information. FLIM analysis can profoundly impact the interpretations of biochemical and physical phenomena.

In time-domain approaches, FLIM usually measures a three-dimensional data cube (x-y-t), obtained with a time-correlated single-photon counting (TCSPC) system [13–16], a time-gated camera [17–21], or a streak camera [22–28]. A time-resolved histogram h_m at (x, y) in a measured data cube can be expressed as:

$$h_m = \sum_{k=0}^{m} irf_{k-m} \cdot f_m + \epsilon_m, \tag{1}$$

$$irf_m = \int_{m\Delta t}^{(m+1)\Delta t} irf(t)dt, f_m = \int_{m\Delta t}^{(m+1)\Delta t} f(t)dt, \; m = 0, \; \dots, M-1, \tag{2}$$

where M is the number of time-bins and Δt is the bin width.

$f(t)$ in Eq. (1) is the underlying fluorescence decay, usually following a multi-exponential decay model,

$$f(t) = A \sum_{p=0}^{P-1} q_p \exp\left(-t/\tau_p\right), \; \sum_{p=0}^{P-1} q_p = 1, \tag{3}$$

where A is the amplitude, q_p and τ_p are the fraction and lifetime of the pth component (p = 0, ... , P-1).

$irf(t)$ is the instrument response function (IRF), often measured using a sample with a much shorter lifetime than the width of the excitation pulse or a scattering solution. The full width at half maximum (FWHM) of IRF, Δt_{IRF}, is given by [16, 29].

$$\Delta t_{\text{IRF}}^2 = \Delta t_{\text{optical}}^2 + \Delta t_{\text{tts}}^2 + \Delta t_{\text{jitter}}^2, \tag{4}$$

where $\Delta t_{\text{optical}}$ is the optical pulse width, Δt_{tts} is the detector transit time spread, and Δt_{jitter} is the detection and timing electronics jitter. With $\boldsymbol{h} = (h_0, \dots, h_{M-1})$ and $\boldsymbol{irf} = (irf_0, \dots, irf_{M-1})$ already measured, A, $\boldsymbol{q} = (q_0, \dots, q_{P-1})$ and $\boldsymbol{\tau} = (\tau_0, \dots, \tau_{P-1})$ can be extracted with a lifetime determination algorithm (LDA).

LDAs can be divided into two categories: fitting and non-fitting methods.

Fitting methods solve a nonlinear minimization problem *argmin* χ^2, where χ^2 is a merit function revealing the goodness of fit. χ^2 is determined by statistics models, such as least square estimation (LSE) or maximum likelihood estimation (MLE), using pixelwise or global fitting modes [30]. Fitting methods suffer from slow analysis due to extensive intrinsic convolutions. The Laguerre expansion method converts the nonlinear-fitting problem to a linear-fitting problem [31, 32], and speeds up deconvolution procedures.

Non-fitting methods can provide lifetime parameters much faster than fitting methods, but some can only provide average lifetimes or graphic representation rather than specific lifetime components. Non-fitting methods should be used carefully according to applications. As discussed in [33–35], the intensity-weighted average lifetime, $\tau_I = \sum_{p=0}^{P-1} q_p \tau_p^2 / \sum_{p=0}^{P-1} q_p \tau_p$, can estimate the average collisional constant k_q from the Stern–Volmer constant K_D, whereas the amplitude-weighted average lifetime, $\tau_A = \sum_{p=0}^{P-1} q_p \tau_p$, can estimate FRET efficiency and assess dynamic quenching behaviors, as described by the Stern–Volmer equation. τ_I was previously misused in FRET efficiency estimation, introducing significant bias when the decay follows a multi-exponential decay model.

The practical use of diverse LDAs depends on applications. Due to the time-consuming estimation procedure, fitting methods are generally used for offline analysis, providing more decay information, including lifetime components and average lifetimes, than most non-fitting methods. Meanwhile, non-fitting methods are suitable for real-time FLIM applications, as they are much faster and more hardware-friendly.

This work attempts to review widely used and newly developed cutting-edge LDAs for time-domain FLIM analysis. The subsequent sections are arranged as follows. Section 2 reviews fitting methods, such as the least squares estimation, maximum likelihood estimation, global fitting using iterative convolution and variable projection approaches, and Laguerre expansion deconvolution methods. Section 3 reviews non-fitting methods, including rapid lifetime determination, center-of-mass, integral extraction, phasor, τ_A/τ_I, deep learning, and histogram clustering methods. Finally, we conclude this review in Section 4 and speculate on future research directions.

2. Fitting methods

Fitting methods use iterative constrained optimization algorithms to estimate fitting parameters based on a specified decay kinetics model. Feedback is provided at each iteration for updating or terminating the process with a criterion. The weighted least squares estimation (LSE) and the maximum likelihood estimation (MLE) are commonly used for FLIM analysis, and these methods have been compared using synthetic and experimental data [36–39]. Although it is well known that MLE is more efficient and accurate than LSE, as FLIM data is Poisson distributed rather than Gaussian distributed [40–44], LSE is still more popular than MLE in FLIM analysis. Attempts have been made to provide robust MLE fitting algorithms, which are counterparts of the Levenberg–Marquardt procedure for LSE [45–47].

The global fitting method can accelerate analysis by assuming spatial lifetime invariances to reduce the degree of freedom significantly. The Laguerre expansion method can accelerate analysis by converting the nonlinear-fitting problem to a linear-fitting problem to speed up deconvolution procedures.

2.1 Weighted least squares estimation (LSE)

The weighted LSE defines a merit function as [48, 49],

$$\chi^2 = \sum_m \frac{\left(h_m - \hat{h}_m\right)^2}{\sigma_m^2}, \tag{5}$$

where h_m is the measured counts, $\hat{h}_m$ is the predicted counts, and σ_m is the count deviation in Bin m. By minimizing χ^2, lifetime parameters can be obtained. For Poisson distributed data, the variance is equal to the actual value ${h_m}^{true}$ of Bin m, that is, $\sigma_m = \sqrt{h_m^{true}}$. As this actual value is not available, χ^2 can be approximated as:

$$\chi_P^2 = \sum_m \frac{\left(h_m - \hat{h}_m\right)^2}{h_m}, \chi_N^2 = \sum_m \frac{\left(h_m - \hat{h}_m\right)^2}{max\,(h_m, 1)}, \tag{6}$$

where χ_P^2 and χ_N^2 are Pearson's and modified Neyman's χ^2. Studies show that Neyman weighting underestimates the target answer, whereas Pearson weighting affords an acceptable answer when the total count is more than 1000 [39].

2.2 Maximum likelihood estimation (MLE)

MLE maximizes the probability that the data can occur given a model and a set of parameters. The likelihood L can be expressed as [50, 51].

$$L = \prod_m P\left(h_m; \hat{h}_m\right), \tag{7}$$

where $P\left(h_m; \hat{h}_m\right)$ is the probability that Bin m has h_m counts if the actual value is $\hat{h}_m$. Because the measurements are Poisson distributed, the Poisson likelihood becomes.

$$L_P = \prod_m \frac{\hat{h}_m^{\,h_m}}{h_m!} e^{-\hat{h}_m}. \tag{8}$$

If Eq. (8) is divided by the maximum possible likelihood $L_P(h_m; h_m)$, then

$$\lambda = L_P\left(h_m; \hat{h}_m\right) / L_P(h_m; h_m). \tag{9}$$

A merit function can be defined as:

$$\chi^2_{\text{mle}} = -2\ln\lambda = 2\sum_m \left(\hat{h}_m - h_m\right) - 2\sum_{m,\, h_m \neq 0} h_m \ln\left(\hat{h}_m / h_m\right). \tag{10}$$

Based on the Poisson likelihood function, MLE can ensure unbiased estimations.

2.3 Global fitting (GF)

The global fitting method uses a least squares estimate χ^2_{GF} for all histograms as the merit function,

$$\chi^2_{GF} = \sum_s \sum_m \frac{\left(h_m^{(s)} - \hat{h}_m^{(s)}\right)^2}{\sigma_m^{(s)2}}, \tag{11}$$

where the superscript s represents Histogram s. GF treats the lifetime components $\boldsymbol{\tau}$ as constants, but the amplitude A and fraction $\boldsymbol{q}$ as variables for all histograms. There are two strategies for implementing GF, the iterative convolution and the variable projection approaches. The variable projection approach appears to be faster than the iterative convolution method, as investigated in [52].

2.3.1 Iterative convolution

The underlying decay is estimated with.

$$\hat{f}_m^{(s)} = \hat{A}^{(s)} \sum_{p=0}^{P-1} \hat{q}_p^{(s)} e^{\frac{-t_m}{\hat{\tau}_p}}, \tag{12}$$

where $\hat{\tau}_p$ are estimated constant lifetimes for all histograms with $\hat{A}^{(s)}$ and $\hat{q}_p^{(s)}$ being the parameters for Histogram s.

Then the estimated signal can be expressed as:

$$\hat{h}_m^{(s)} = \sum_{k=0}^{m} irf_{k-m} \cdot \hat{f}_m^{(s)},\ m = 0,\ \dots,\ M-1. \tag{13}$$

Using Eq. (13), we can minimize Eq. (11) with constrained LSE. The analysis speed is significantly affected by chosen initial conditions. S. Pelet *et al.* compared different strategies for initial guesses [30] and proposed an efficient image segmentation method.

2.3.2 Variable projection

The idea of global fitting with variable projection is to minimize a projection function that depends only on nonlinear parameters $\boldsymbol{\tau}$ and obtain linear parameters $A^{(s)}$ and $\boldsymbol{q}^{(s)}$. A matrix whose columns only depend on $\boldsymbol{\tau}$ is constructed,

$$\boldsymbol{\Phi}(\hat{\boldsymbol{\tau}}) = [\boldsymbol{\varphi}_0(\hat{\tau}_0),\ \dots,\ \boldsymbol{\varphi}_{P-1}(\hat{\tau}_{P-1})], \tag{14}$$

where $\boldsymbol{\varphi}_p(\hat{\tau}_p) = \left[\varphi_p(\hat{\tau}_p;t_0),\ \dots,\ \varphi_p(\hat{\tau}_p;t_{M-1})\right]^T$, $\varphi_p(\hat{\tau}_p;t_m) = \sum_{k=0}^{m} irf_{k-m} \cdot \exp\left(-\frac{t_m}{\hat{\tau}_p}\right)$, and p = 0, ..., P-1.

Then the estimated signal can be written as:

$$\hat{h}_m^{(s)} = \sum_{p=0}^{P-1} \hat{a}_p^{(s)} \varphi_p(\hat{\tau}_p;t_m), \tag{15}$$

where $\hat{a}_p{}^{(s)} = \hat{A}^{(s)} \hat{q}_p{}^{(s)}$. Eq. (11) becomes.

$$\chi_{GF}^2 = \sum_s \sum_m \frac{\left(h_m^{(s)} - \sum_{p=0}^{P-1} \hat{a}_p^{(s)} \varphi_p(\hat{\tau}_p;t_m)\right)^2}{\sigma_m^{(s)2}}, \tag{16}$$

$$\chi_{GF}^2 = \sum_s \|\overline{\boldsymbol{h}}^{(s)} - \overline{\boldsymbol{\Phi}}(\hat{\boldsymbol{\tau}})\hat{\boldsymbol{a}}^{(s)}\|^2, \tag{17}$$

where $\overline{\boldsymbol{h}}^{(s)}$ is $\boldsymbol{h}^{(s)}$ weighted by $\sigma_m^{(s)}$, $\overline{\boldsymbol{\Phi}}(\hat{\boldsymbol{\tau}})$ is $\boldsymbol{\Phi}(\hat{\boldsymbol{\tau}})$ weighted by $\sigma_m^{(s)}$, and $\hat{\boldsymbol{a}}^{(s)} = \left[\hat{a}_0{}^{(s)},\ \dots,\ \hat{a}_{P-1}{}^{(s)}\right]^T$.

For a given $\hat{\boldsymbol{\tau}}$, Eq. (17) is minimized when $\hat{\boldsymbol{a}}^{(s)} = \overline{\boldsymbol{\Phi}}(\hat{\boldsymbol{\tau}})^{-}\overline{\boldsymbol{h}}^{(s)}$, where $\overline{\boldsymbol{\Phi}}(\hat{\boldsymbol{\tau}})^{-}$ is the symmetric generalized inverse of $\overline{\boldsymbol{\Phi}}(\hat{\boldsymbol{\tau}})$. Then Eq. (17) can be rewritten as:

$$\chi_{GF}^2 = \sum_s \|\boldsymbol{P}_{\boldsymbol{\Phi}(\hat{\boldsymbol{\tau}})}^{\perp} \overline{\boldsymbol{h}}^{(s)}\|^2, \tag{18}$$

where $\boldsymbol{P}_{\boldsymbol{\Phi}(\hat{\boldsymbol{\tau}})}^{\perp} = \boldsymbol{I} - \overline{\boldsymbol{\Phi}}(\hat{\boldsymbol{\tau}})\overline{\boldsymbol{\Phi}}(\hat{\boldsymbol{\tau}})^{-}$ [52], which can be calculated by matrix decomposition of $\overline{\boldsymbol{\Phi}}(\hat{\boldsymbol{\tau}})$ using the QR method. Eq. (18) only has the nonlinear parameter, reducing the minimization parameter space considerably. Once $\hat{\boldsymbol{\tau}}$ is obtained by minimizing Eq. (18), the linear parameter $\hat{\boldsymbol{a}}^{(s)}$ can be obtained as a solution of $\boldsymbol{h}^{(s)} = \boldsymbol{\Phi}(\hat{\boldsymbol{\tau}})\hat{\boldsymbol{a}}^{(s)}$. The implementation can be achieved with the VARP2 code [53].

2.4 Laguerre expansion (LE)

Due to the IRF introduced in FLIM data, iterative fitting methods include an enormous amount of convolutions, which is time-consuming [54]. Furthermore, iterative deconvolution methods require acquiring a considerable number of counts, which would increase the acquisition time. Numerous mathematical tools have been devised for deconvolution [55–59], and LE is faster and more robust than others. LE estimates the underlying fluorescence decay $\boldsymbol{f}$ with an ordered set of discrete-time Laguerre basis functions (LBFs) [60, 61],

$$\hat{f}_m = \sum_{l=0}^{L-1} \hat{c}_l b_l(m;\alpha), \tag{19}$$

where L and α are the basic parameters and $\hat{c}_l$ is the estimated l^{th} expansion coefficient. The lth discrete-time LBF is defined as:

$$b_l(m;\alpha) = \alpha^{\frac{(m-l)}{2}}(1-\alpha)^{\frac{1}{2}}\sum_{i=0}^{l}(-1)^i\binom{m}{i}\binom{l}{i}\alpha^{l-i}(1-\alpha)^i, \tag{20}$$

where $0 < \alpha < 1$ and $l = 0, \ldots, L\text{-}1$.

Substituting Eq. (19) for f_m in Eq. (1), the estimated signal becomes:

$$\hat{h}_m = \sum_{i=0}^{m}\sum_{l=0}^{L-1}\hat{c}_l \cdot irf_{m-i} \cdot b_l(i;\alpha) = \sum_{l=0}^{L-1}\hat{c}_l \cdot v_l(m;\alpha), \tag{21}$$

where $v_l(m;\alpha) = \sum_{i=0}^{m} irf_{m-i} \cdot b_l(i;\alpha)$. Then, Eq. (5) becomes.

$$\chi^2_{LE} = \sum_m \frac{\left(h_m - \sum_{l=0}^{L-1}\hat{c}_l \cdot v_l(m;\alpha)\right)^2}{\sigma_m^2}, \tag{22}$$

where $\boldsymbol{V} = [\boldsymbol{v}_0, \ldots, \boldsymbol{v}_{L\text{-}1}]$ and $\boldsymbol{v}_l = [v_l(0,\alpha), \ldots, v_l(M\text{-}1;\alpha)]^T$. Eq. (22) can be minimized with the ordinary and constrained LSE, as demonstrated in [61]. Once $\hat{\boldsymbol{c}} = [\hat{c}_0, \ldots, \hat{c}_{L-1}]^T$ is determined, $\hat{\boldsymbol{f}}$ can be recovered with Eq. (19). Then decay parameters can be extracted from $\hat{\boldsymbol{f}}$ using the fitting methods mentioned above or the following non-fitting methods.

Setting proper L and α depends on the lifetime dynamic range and the measurement window $T = M\Delta t$ [32]. Automated Laguerre deconvolution methods have been reported to optimize L and α during the deconvolution routine [62, 63].

3. Non-fitting methods

Fitting methods usually require experienced users to set proper initial conditions and require time-consuming iterative computations, making unsupervised non-fitting methods attractive for robust, real-time FLIM analysis. We review the rapid lifetime determination, center-of-mass, integral extraction, phasor, τ_A/τ_I, deep learning, and histogram clustering methods.

3.1 Rapid lifetime determination (RLD)

RLD needs two time-gated signals for the lifetime estimation of mono-exponential decays. The initial RLD method utilizes two consecutive time-gates of equal widths [64], as shown in **Figure 1(a)**. By integrating the signal within the time-gated windows, the lifetime can be determined using,

$$\tau_{RLD} = \frac{-\Delta t_{TG}}{ln\,(D_1/D_0)}, \tag{23}$$

where D_0 and D_1 denote the integrated signal in Gates 1 and 2, and Δt_{TG} denotes the time-gate width. This calculation is simple and fast; however, it has limited estimation accuracy and precision in a narrow lifetime range determined by the gate width. To optimize RLD, several strategies with overlapping and unequal time gates, as shown in **Figure 1(b)**, have been proposed [65–67], suggesting optimized gating schemes. In optimized RLD schemes, the ratio of D_1 and D_0 can be expressed as:

$$DR = \frac{D_1}{D_0} = \frac{exp\left(-Y\frac{\Delta t_{TG}}{\tau}\right) - exp\left[-(Y+W)\frac{\Delta t_{TG}}{\tau}\right]}{1 - exp\left(-\frac{\Delta t_{TG}}{\tau}\right)}, \tag{24}$$

where Δt_{TG} is the width of Gate 1 and Y and W are the rising edge and width coefficient of Gate 2. Lifetimes τ_{RLD} can be solved by applying Newton's method.

RLD can deal with multiple types of FLIM data, including TCSPC, streak camera, time-gated counting, and time-gated imaging data. However, RLD performs differently for uncorrelated and correlated datasets, as analyzed in ref. [66]. For multi-exponential decays, RLD can provide average lifetimes which are neither intensify- nor amplitude-weighted. Additionally, with three-time gates, the lifetime estimation dynamic range of RLD can be expanded for mono-exponential decays [21]. With four time-gates, we can extract lifetime components of bi-exponential decays [68, 69]. IRF is usually neglected in RLD, introducing estimation bias, especially for short lifetimes comparable with IRF's width.

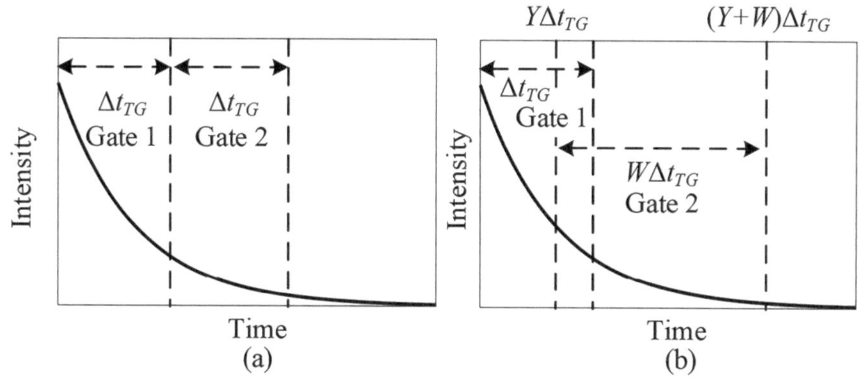

Figure 1.
Rapid lifetime determination (RLD) schemes with (a) equal and (b) unequal time-gate widths.

3.2 Center-of-mass method (CMM)

CMM provides intensity-weighted average lifetimes for multi-exponential decays [70–72],

$$\tau_{CMM} = \frac{\int_0^\infty t \cdot h(t)dt}{\int_0^\infty h(t)dt} - \frac{\int_0^\infty t \cdot irf(t)dt}{\int_0^\infty irf(t)dt} = \frac{\sum_{p=0}^{P-1} q_p \tau_p^2}{\sum_{p=0}^{P-1} q_p \tau_p} \approx \frac{\sum_{m=0}^{M-1} t_m \cdot h_m}{\sum_{m=0}^{M-1} h_m} - \frac{\sum_{m=0}^{M-1} t_m \cdot irf_m}{\sum_{m=0}^{M-1} irf_m}. \tag{25}$$

Li et al. proposed two versions of bi-exponential CMM (BCMM) that provide lifetime components information [73].

3.3 Integral extraction method (IEM)

For IEM, deconvolution is required to obtain $\hat{f}$ with which the average lifetime can be determined as [74, 75].

$$\tau_{IEM} = -\frac{\int_0^\infty g(t)dt}{\int_0^\infty g'(t)dt} \approx \sum_{p=0}^{P-1} q_p \tau_p \approx -\frac{\sum_{m=0}^{M-1} S_m \cdot \hat{f}_m}{\sum_{m=0}^{M-1} \frac{\hat{f}_m - \hat{f}_{m-1}}{\Delta t}} = -\frac{\Delta t \sum_{m=0}^{M-1} S_m \cdot \hat{f}_m}{\hat{f}_{M-1} - \hat{f}_0}, \tag{26}$$

where S_m = [1/3, 4/3, 2/3, … , 4/3, 1/3] are the coefficients for numerical integration based on Simpson's rule,

$$g(t) = A \sum_{p=0}^{P-1} q_p \tau_p e^{-\frac{t}{\tau_p}} \left[1 - e^{-\frac{\Delta t}{\tau_p}}\right]. \tag{27}$$

τ_{IEM} is an estimator of amplitude-weighted lifetimes.

3.4 Phasor method

The phasor method transforms each histogram into a phasor, like a vector. The sine-cosine transforms of decays are represented in a phasor plot as a two-dimensional histogram [76–80]. Phasor components g and s for time-domain FLIM can be expressed as:

$$g = \frac{\int_0^\infty f(t) \cdot \cos(\omega t)dt}{\int_0^\infty f(t)dt} = \frac{\sum_{p=0}^{P-1} \frac{q_p \tau_p}{1+\omega^2 \tau_p^2}}{\sum_{p=0}^{P-1} q_p \tau_p} = \frac{Re + s \bullet B}{A}, \tag{28}$$

$$s = \frac{\int_0^\infty f(t) \cdot \sin(\omega t)dt}{\int_0^\infty f(t)dt} = \frac{\sum_{p=0}^{P-1} \frac{\omega q_p \tau_p^2}{1+\omega^2 \tau_p^2}}{\sum_{p=0}^{P-1} q_p \tau_p} = \frac{A \bullet Im - B \bullet Re}{A^2 + B^2}, \tag{29}$$

where

$$Re = \frac{\int_0^\infty h(t) \cdot \cos(\omega t) dt}{\int_0^\infty h(t) dt} \approx \frac{\sum_{m=0}^{M-1} h_m \cdot \cos(\omega t_m)}{\sum_{m=0}^{M-1} h_m}, \tag{30}$$

$$Im = \frac{\int_0^\infty h(t) \cdot \sin(\omega t) dt}{\int_0^\infty h(t) dt} \approx \frac{\sum_{m=0}^{M-1} h_m \cdot \sin(\omega t_m)}{\sum_{m=0}^{M-1} h_m}, \tag{31}$$

$$A = \frac{\int_0^\infty irf(t) \cdot \cos(\omega t) dt}{\int_0^\infty irf(t) dt} \approx \frac{\sum_{m=0}^{M-1} irf_m \cdot \cos(\omega t_m)}{\sum_{m=0}^{M-1} irf_m}, \tag{32}$$

$$B = \frac{\int_0^\infty irf(t) \cdot \sin(\omega t) dt}{\int_0^\infty irf(t) dt} \approx \frac{\sum_{m=0}^{M-1} irf_m \cdot \sin(\omega t_m)}{\sum_{n=0}^{M-1} irf_m}. \tag{33}$$

Phasor plots' interpretation is usually user-dependent by manually selecting regions of interest in a phasor plot to find corresponding regions in fluorescence images. Based on the feature that pixels with similar fluorescence decays tend to congregate and form a cluster in a phasor plot, machine learning techniques have been developed with clustering methods [81] to automatically organize phasors into sensible groupings. Up to four-lifetime components can be resolved from a phasor plot using the rule of a linear combination of phasors [82], a graphical approach [83], and a computational nonlinear minimization algorithm [83].

A weighted average lifetime, τ_{Phasor}, can also be derived using phasors,

$$\tau_{\text{Phasor}} = \frac{s}{g\omega} = \frac{\sum_{p=0}^{P-1} \frac{q_p \tau_p^2}{1+\omega^2 \tau_p^2}}{\sum_{p=0}^{P-1} \frac{q_p \tau_p}{1+\omega^2 \tau_p^2}}, \tag{34}$$

where $\omega = 2\pi/T, T = M\Delta t$ is the measurement window. The weights of τ_{Phasor} are $\frac{q_p \tau_p}{1+\omega^2 \tau_p^2}$. If $\tau_p << T$, then the weights are approximately equal to $q_p \tau_p$, i.e. τ_{Phasor} is close to intensity-weighted lifetimes.

3.5 τ_A/τ_I method

We proposed the τ_A/τ_I method, a multi-exponential decay visualization method using two types of average lifetimes, τ_I and τ_A. As mentioned previously (discussed in ref. [35]), a fluorescence decay can be approximated by a bi-exponential decay model, so that the ratio of τ_A and τ_I can be expressed as:

$$\frac{\tau_A}{\tau_I} = \frac{(1 + q_1(R-1))^2}{1 + q_1(R^2 - 1)}, \tag{35}$$

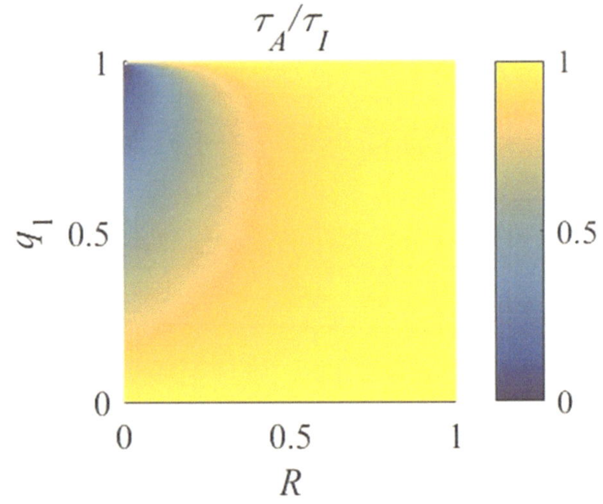

Figure 2.
Distribution of τ_A/τ_I *with* $q_1 = 0 \sim 1$ *and* $R = 0 \sim 1$ *[35].*

where $R = \tau_1/\tau_2$. The distribution of τ_A/τ_I (**Figure 2**) shows that when $R \approx 1$ or $q_1 \approx 0$ or 1, $\tau_A/\tau_I \approx 1$. With a decrease in R or an increase of q_1, τ_A/τ_I decreases. Therefore, the ranges of q_1 and R of a pixel can be determined by τ_A/τ_I.

With a TCSPC dataset of tSA201 cells, **Figures 3** and **4** show the results of the selected pixels within different τ_A/τ_I ranges in (a) τ_I, (b) τ_A, (c) τ_A/τ_I images, (d) histograms, (e) phasor plots, and (f) τ_A/τ_I plots. For the pixels within $\tau_A/\tau_I = 0.2 \sim 0.5$ (**Figure 3**), the histograms clearly show that τ_A is much smaller than τ_I, which means the difference between τ_1 and τ_2 is significant. **Figure 3** (f) shows that the ranges of q_1 and R are approximately $0.5 \sim 1$ and $0 \sim 0.2$, respectively. For the pixels with $\tau_A/\tau_I = 0.5 \sim 1$ (**Figure 4**), τ_A is closer to τ_I, meaning the pixels have decays close to mono-exponential.

τ_A/τ_I is an intuitive tool for visualizing multi-exponential decays in a lifetime image. Separating the average lifetime images with τ_A/τ_I is easier than phasor plots because τ_A/τ_I is one-dimensional, and phasors are two-dimensional. Furthermore, τ_A/τ_I can intuitively show the q_1 and R ranges.

3.6 Deep learning (DL)

Recently, DL-based FLIM analysis methods have been reported. DL features hierarchical representation learning by extracting high-level features through multiple nonlinear transformations of low-level features. DL shows a powerful ability to learn complex data and functions. The advent of DL breaks the conventional "model-driven" paradigm and offers us a new "data-driven" approach to solving general optimization problems. Given sufficient labeled training data, DL algorithms can directly map the raw input data to their corresponding results, thus avoiding time-consuming iterative optimization processes. Wu et al., first employed DL with a multi-layer perception (MLP) network using one input layer, one output layer, and two hidden layers [84]. The input layer has 57 entries depending on the number of time-bins in a histogram. The output layer has four neurons mapping four-lifetime parameters for bi-exponential decays. The investigation reveals that MLP can provide

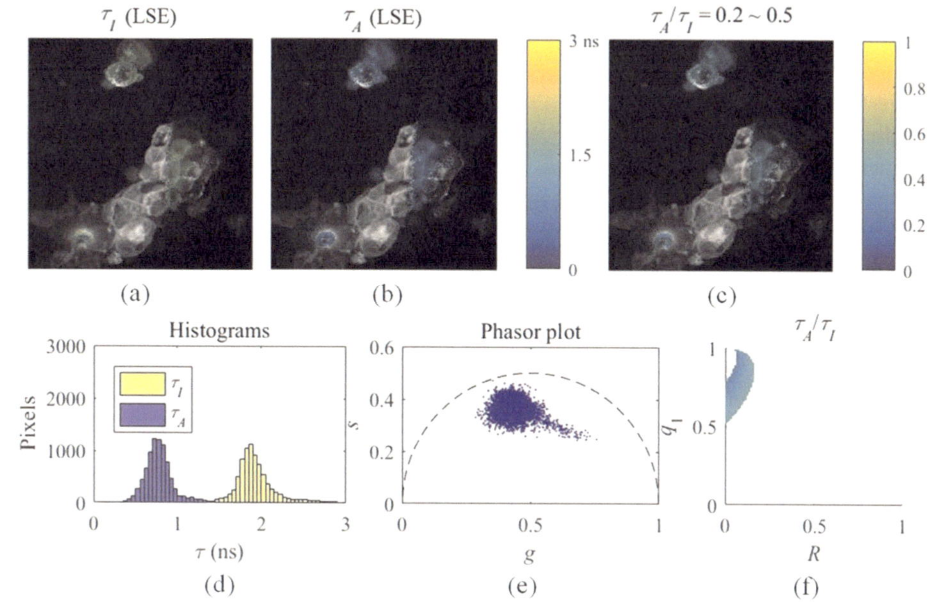

Figure 3.
(a) τ_I-intensity image, (b) τ_A-intensity image, (c) τ_A/τ_I ratio image, (d) histograms of τ_I (yellow) and τ_A (blue), (e) phasor plot, and (f) distribution of τ_A/τ_I of the selected pixels in (c) with τ_A/τ_I = 0.2 ~ 0.5 [35].

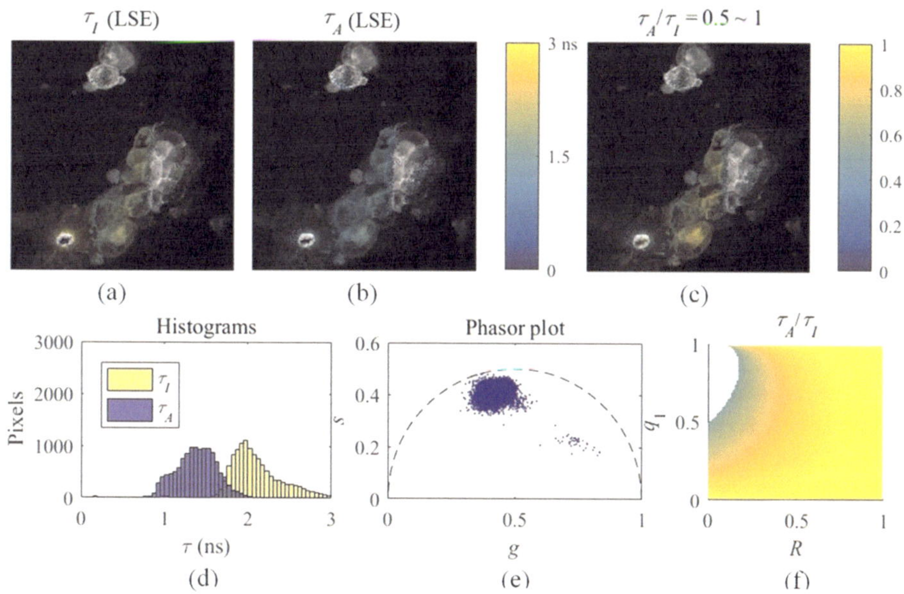

Figure 4.
(a) τ_I-intensity image, (b) τ_A-intensity image, (c) τ_A/τ_I ratio image, (d) histograms of τ_I (yellow) and τ_A (blue), (e) phasor plot, and (f) distribution of τ_A/τ_I of the selected pixels in (c) with τ_A/τ_I = 0.5 ~ 1 [35].

comparable or better performances and generate lifetime images at least 180-fold faster than conventional LSE. However, IRF was not considered in their work. Smith et al. proposed a 3D convolutional neural network (CNN) architecture named fluorescence lifetime imaging network (FLI-Net) to quantify fluorescence decays at fast speeds [85]. FLI-Net is designed and trained for TCSPC FLIM, and gated intensity charged-coupled device (ICCD) based FLIM. FLI-Net's input is a 3D data cube (x, y, t), and output is bi-exponential decay parameters. Synthetic data is used for training, avoiding acquiring massive training datasets experimentally. FLI-Net is about 30-fold faster than SPCImage, the widely used FLIM processing software. FLI-Net outperforms LSE and MLP, especially with low photon counts range from 25 to 100. Xiao et al. proposed an easier and faster trained 1D CNN architecture named 1D-ConvResNet [86]. Compared with 2D or 3D CNNs, 1D-ConvResNet is more hardware-friendly and can be implemented on field-programmable gate array (FPGA) devices. Synthetic data is used for training and the training time with a CPU (Intel i7–4790) is about 0.5 hours, 8-fold faster than FLI-Net with a GPU (NVIDIA TITAN Xp GPU). Experimental FLIM datasets with an intensity threshold of 100 counts per pixel were used for validation, and analyzing a 256 × 256 image takes several seconds on a laptop. To generate high-quality FLIM images under photon-starved conditions (50 counts per pixel), Chen et al. introduced a method called flimGANE [87]. It was derived from the Wasserstein GAN algorithm, where an "artificial" high-photon-count fluorescence decay histogram can be produced with a generator with a low-photon-count input. Using a well-trained generator and an estimator, a low-quality decay histogram can be mapped to a high-quality counterpart and provides bi-exponential lifetime parameters within 0.32 ms/pixel using a CPU, which is 258-fold faster than LSE and 2800-fold faster than MLE in generating a 512 × 512 FLIM image. However, it takes up to 500 h to fully train the network. Additionally, using time-resolved single-pixel datasets, a deep CNN named Net-FLICS (fluorescence lifetime imaging with compressed sensing) was reported [88].

DL methods for spatial resolution enhancement of FLIM images are also developed, including SRI-FLIMnet for reconstructing high-resolution images from low-resolution 3D FLIM data [89] and CNN-based denoising method removing noise in phasor plots after the K-means clustering segmentation [90].

Although DL is promising in real-time FLIM analysis even under starve photon-starved conditions and needs no user-dependent initial conditions, it suffers from long training times (hours to days) when retaining is required due to the change of IRF. Whenever the laser source for excitation and the detector are changed, the IRF can also vary, and the network should be retrained, reducing the universality. Recently, many attempts have been made to alleviate this disadvantage. For example, Zang proposed an online neural network training method using the extreme learning machine (ELM) [91]. ELM does not require a back-propagation process in the training phase. Therefore, it provides a much faster training speed, enabling online network training for any system configuration. The phasor coordinates of a fluorescence decay can also serve as the network's inputs, which can be viewed as a simple feature engineering to reduce the training time significantly [92]. Despite some trade-offs in current DL algorithms, it is no doubt that DL has excellent potential in a wide range of FLIM-related applications. DL algorithms can also be implemented on edge-computing platforms like FPGAs and smartphones to develop intelligent and portable FLIM devices [93].

3.7 Histogram clustering (HC)

The HC method devised by Li et al. can improve FLIM analysis speed and accuracy by sorting histograms with similar profiles in a dataset into several clusters and significantly reducing the number of histograms to be analyzed [94]. HC implements

LDA	t_{exe} (s)	*MSE*				
		q_1	τ_1 (ns^2)	τ_2 (ns^2)	τ_A (ns^2)	τ_I (ns^2)
Without HC						
LE-LSE	389.45	0.019	0.173	0.198	0.110	0.027
LE-IEM	62.30	X	X	X	0.102	X
CMM	0.20	X	X	X	X	0.185
IC	724.82	0.100	X	X	0.678	1.098
VP	3.34	0.033	X	X	0.122	0.178
With HC						
LE-LSE	3.36	0.011	0.102	0.104	0.037	0.025
LE-IEM	0.63	X	X	X	0.038	X
CMM	0.20	X	X	X	X	0.180
IC	11.85	0.017	X	X	0.102	0.050
VP	0.31	0.014	X	X	0.048	0.093

Table 1.
t_{exe} and MSE evaluated by lifetime determination algorithms without and with histogram clustering [94].

clustering with two features of a histogram. It is a preprocessing method that can work with the LDAs mentioned above. The performances for producing decay parameter images without and with HC using synthetic and experimental datasets were investigated [94]. The execution time t_{exe} and the mean squared error (MSE) of a FLIM dataset following a bi-exponential decay model with 150 × 150 pixels and 256 time-bins are shown in **Table 1**. LE-LSE and LE-IEM stand for LSE and IEM with the Laguerre expansion method; IC and VP represent iterative convolution and variable expansion global fitting approaches.

For different output types, we suggested the fastest FLIM analysis methods: 1) LE-LSE with HC for all lifetime component images with t_{exe} = 3.36 s, 116-fold shorter than t_{exe} without HC; 2) VP with HC for constant lifetimes, q_1, τ_A, and τ_I images with t_{exe} = 0.31 s, 10-fold shorter than t_{exe} without HC; and 3) LE-IEM with HC as the second choice for τ_A with t_{exe} = 0.63 s, 98-fold shorter than t_{exe} without HC, and CMM as the second choice for τ_I with t_{exe} = 0.2 s without or with HC (biased if the most significant lifetime > $T/4$). The analysis was conducted in MATLAB, and it can be translated to C or other environments to speed up the analysis.

HC can benefit applications demanding real-time FLIM, such as clinical diagnosis and fast screening. In the future, deep learning methods can be employed for unsupervised histogram clustering.

4. Conclusions

Lifetime determination algorithms (LDAs) play a vital role in FLIM analysis. The results provided by diverse LDAs profoundly impact the interpretation of observed phenomena. Numerous approaches have been developed with notable features in terms of photon efficiency, estimation speed, accuracy, and precision. Fast FLIM analysis under low-photon conditions is universally desirable, especially for real-time and single-molecular FLIM. This mini review focuses on the popular and cutting-edge LDAs used in time-domain FLIM. The progress in LDAs has shown possibilities for real-time FLIM under low photon conditions.

Fitting methods estimate lifetime parameters by solving an iterative minimization problem. The least squares estimation (LSE) introduces bias even with high photon counts (>1000), while the maximum likelihood estimation (MLE) is unbiased and photon efficient, as FLIM data is Poisson distributed. With robust algorithms, MLE instead of LSE can be used as a standard. The iterative convolution of IRF with exponential decays in LSE and MLE is time-consuming. The global fitting and Laguerre expansion methods have been proposed to accelerate analysis.

Non-fitting methods are attractive, as they can be much faster and more hardware-friendly than fitting methods because of simple calculations. However, non-fitting methods for a specific application should be careful. RLD using two time-gates is suitable for mono-exponential decays but may introduce a bias for lifetimes comparable with IRF's width. CMM and IEM deliver intensity- and amplitude-weighted average lifetimes, respectively. The phasor method can graphically show exponential decays in a phasor plot. The τ_A/τ_I method is a multi-exponential decay visualization method that uses two types of average lifetimes, which can intuitively show lifetime components' fraction and lifetime ratio. The deep learning method employs fully connected networks, 3D CNN, 1D CNN, and GAN algorithms in rapid FLIM analysis under photon-starved conditions, down to 50 counts per pixel. However, its long retraining time (hours to days) limits its universality when IRF is changed. The histogram clustering (HC) method suggests that histograms having similar profiles can be clustered for one calculation. By embedding HC, the algorithms mentioned above can be significantly accelerated with an execution time down to 30 μs/pixel.

In summary, the existing LDAs reveal different features in FLIM analysis, and it is hard to judge which one is replaceable. We believe that software with multiple embedded LDAs is desirable for end-users, providing comprehensive and complementary information for robust analysis. In addition, the newly developed "data-driven" DL-based LDAs and LDA accelerating method HC are promising, as they can not only enhance speed but also accuracy, especially for photon-starved applications. More efforts are expected to shrink retraining time for DL and increase clustering performance for HC via deep learning.

Acknowledgements

Supported by the Scientific Instrument Developing Project of the Chinese Academy of Sciences (GJJSTD20220006), Science Foundation of the Chinese Academy of Sciences (CXJJ-21S006), China Scholarship Council, Biotechnology and Biological Sciences Research Council (BB/V019643/1) and Medical Research Scotland.

Conflict of interest

The authors declare no conflict of interest.

Abbreviations

FLIM	Fluorescence Lifetime Imaging
FRET	Förster Resonance Energy Transfer
IRF	Instrument Response Function
TCSPC	Time-Correlated Single-Photon Counting
LDA	Lifetime Determination Algorithm
LSE	Least Squares Estimation
MLE	Maximum Likelihood Estimation
GF	Global Fitting
LE	Laguerre Expansion
LBF	Laguerre Basis Function
RLD	Rapid Lifetime Determination
CMM	Center-of-Mass Method
IEM	Integral Extraction Method
DL	Deep Learning
HC	Histogram Clustering
MLP	Multi-Layer Perception
CNN	Convolutional Neural Network
GAN	Generative Adversarial Network
ELM	Extreme Learning Machine
FPGA	Field-Programmable Gate Array

Author details

Yahui Li[1]*, Lixin Liu[2], Dong Xiao[3], Hang Li[1], Natakorn Sapermsap[4], Jinshou Tian[1], Yu Chen[4] and David Day-Uei Li[3]

1 Key Laboratory of Ultra-fast Photoelectric Diagnostics Technology, Xi'an Institute of Optics and Precision Mechanics, Xi'an, Shaanxi, China

2 Xidian University, School of Optoelectronic Engineering, Xi'an Shaanxi, China

3 Department of Biomedical Engineering, University of Strathclyde, Glasgow, United Kingdom

4 Department of Physics, University of Strathclyde, Glasgow, United Kingdom

*Address all correspondence to: liyahui@opt.ac.cn

References

[1] Lakowicz JR. Principles of Fluorescence Spectroscopy. 3rd ed. New York: Springer; 2006

[2] Priessner M, Summers PA, Lewis BW, Sastre M, Ying L, Kuimova MK, et al. Selective detection of Cu $^{+}$ ions in live cells via fluorescence lifetime imaging microscopy. Angewandte Chemie, International Edition. 2021;**60**:23148-23153. DOI: 10.1002/anie.202109349

[3] Gadella TWJ, editor. FRET and FLIM techniques. 1st ed. Amsterdam ; Boston: Elsevier; 2009

[4] Wallrabe H, Periasamy A. Imaging protein molecules using FRET and FLIM microscopy. Current Opinion in Biotechnology. 2005;**16**:19-27. DOI: 10.1016/j.copbio.2004.12.002

[5] Bücherl CA, Bader AN, Westphal AH, Laptenok SP, Borst JW. FRET-FLIM applications in plant systems. Protoplasma. 2014;**251**:383-394. DOI: 10.1007/s00709-013-0595-7

[6] Summers PA, Lewis BW, Gonzalez-Garcia J, Porreca RM, Lim AHM, Cadinu P, et al. Visualizing G-quadruplex DNA dynamics in live cells by fluorescence lifetime imaging microscopy. Nature Communications. 2021;**12**:162. DOI: 10.1038/s41467-020-20414-7

[7] Levchenko SM, Pliss A, Peng X, Prasad PN, Qu J. Fluorescence lifetime imaging for studying DNA compaction and gene activities. Light Science Applications. 2021;**10**:224. DOI: 10.1038/s41377-021-00664-w

[8] Pliss A, Zhao L, Ohulchanskyy TY, Qu J, Prasad PN. Fluorescence lifetime of fluorescent proteins as an intracellular environment probe sensing the cell cycle progression. ACS Chemical Biology. 2012;**7**:1385-1392. DOI: 10.1021/cb300065w

[9] Fornasiero EF, Mandad S, Wildhagen H, Alevra M, Rammner B, Keihani S, et al. Precisely measured protein lifetimes in the mouse brain reveal differences across tissues and subcellular fractions. Nature Communications. 2018;**9**:4230. DOI: 10.1038/s41467-018-06519-0

[10] Hirmiz N, Tsikouras A, Osterlund EJ, Richards M, Andrews DW, Fang Q. Highly multiplexed confocal fluorescence lifetime microscope designed for screening applications. IEEE Journal of Selected Topics in Quantum Electronics. 2021;**27**:1-9. DOI: 10.1109/JSTQE.2020.2997834

[11] Lukina M, Yashin K, Kiseleva EE, Alekseeva A, Dudenkova V, Zagaynova EV, et al. Label-free macroscopic fluorescence lifetime imaging of brain tumors. Frontiers in Oncology. 2021;**11**:666059. DOI: 10.3389/fonc.2021.666059

[12] Chen J, Han G, Liu Z, Wang H, Wang D, Zhao J, et al. Recovery mechanism of endoplasmic reticulum revealed by fluorescence lifetime imaging in live cells. Analytical Chemistry. 2022;**94**:5173-5180. DOI: 10.1021/acs.analchem.2c00216

[13] Becker W. Advanced Time-Correlated Single Photon Counting Techniques. Berlin ; New York: Springer; 2005

[14] Hirvonen LM, Suhling K. Wide-field TCSPC: Methods and applications. Measurement Science and Technology.

2017;**28**:012003. DOI: 10.1088/1361-6501/28/1/012003

[15] Liu X, Lin D, Becker W, Niu J, Yu B, Liu L, et al. Fast fluorescence lifetime imaging techniques: A review on challenge and development. Journal of Innovative Optical Health Sciences. 2019;**12**:1930003. DOI: 10.1142/s1793545819300039

[16] Hirvonen LM, Suhling K. Fast timing techniques in FLIM applications. Frontiers of Physics. 2020;**8**:161. DOI: 10.3389/fphy.2020.00161

[17] Dowling K, Hyde SCW, Dainty JC, French PMW, Hares JD. 2-D fluorescence lifetime imaging using a time-gated image intensifier. Optics Communication. 1997; **135**:27-31. DOI: 10.1016/S0030-4018(96)00618-9

[18] Elangovan M, Day RN, Periasamy A. Nanosecond fluorescence resonance energy transfer-fluorescence lifetime imaging microscopy to localize the protein interactions in a single living cell: NANOSECOND FRET-FLIM MICROSCOPY. Journal of Microscopy. 2002;**205**:3-14. DOI: 10.1046/j.0022-2720.2001.00984.x

[19] Grant DM, McGinty J, Ewan J. McGhee, McGhee EJ, Bunney TD, Owen DM, et al. High speed optically sectioned fluorescence lifetime imaging permits study of live cell signaling events. Optics Express 2007;15: 15656–15673. DOI: 10.1364/oe.15.015656.

[20] Ehn A, Johansson O, Arvidsson A, Aldén M, Bood J. Single-laser shot fluorescence lifetime imaging on the nanosecond timescale using a dual image and modeling evaluation algorithm. Optics Express. 2012;**20**:3043-3056. DOI: 10.1364/oe.20.003043

[21] Li Y, Jia H, Chen S, Tian J, Liang L, Yuan F, et al. Single-shot time-gated fluorescence lifetime imaging using three-frame images. Optics Express. 2018;**26**:17936-17947. DOI: 10.1364/OE.26.017936

[22] Fleming G, Morris J, Robinson H. Picosecond fluorescence spectroscopy with a streak camera. Australian Journal of Chemistry. 1977;**30**:2337. DOI: 10.1071/CH9772337

[23] Krishnan RV, Biener E, Zhang J, Heckel R, Herman B. Probing subtle fluorescence dynamics in cellular proteins by streak camera based fluorescence lifetime imaging microscopy. Applied Physics Letters. 2003;**83**:4658-4660. DOI: 10.1063/1.1630154

[24] Biskup C, Zimmer T, Benndorf K. FRET between cardiac Na+ channel subunits measured with a confocal microscope and a streak camera. Nature Biotechnology. 2004;**22**:220-224. DOI: 10.1038/nbt935

[25] Faust S, Dreier T, Schulz C. Temperature and bath gas composition dependence of effective fluorescence lifetimes of toluene excited at 266nm. Chemical Physics. 2011;**383**:6-11. DOI: 10.1016/j.chemphys.2011.03.013

[26] Camborde L, Jauneau A, Brière C, Deslandes L, Dumas B, Gaulin E. Detection of nucleic acid–protein interactions in plant leaves using fluorescence lifetime imaging microscopy. Nature Protocols. 2017;**12**: 1933-1950. DOI: 10.1038/nprot.2017.076

[27] Qu J, Liu L, Chen D, Lin Z, Xu G, Guo B, et al. Temporally and spectrally resolved sampling imaging with a specially designed streak camera. Optics Letters. 2006;**31**:368. DOI: 10.1364/OL.31.000368

[28] Liu L, Qu J, Lin Z, Wang L, Fu Z, Guo B, et al. Simultaneous time- and spectrum-resolved multifocal multiphoton microscopy. Applied Physics B: Lasers and Optics. 2006;**84**: 379-383. DOI: 10.1007/s00340-006-2314-y

[29] Birch DJS, Imhof RE. Time-domain fluorescence spectroscopy using time-correlated single-photon counting. In: Lakowicz JR, editor. Topics in Fluorescence Spectroscopy. Boston: Kluwer Academic Publishers; 2002. pp. 1-95. DOI: 10.1007/0-306-47057-8_1

[30] Pelet S, Previte MJ, Laiho LH, So PT. A fast global fitting algorithm for fluorescence lifetime imaging microscopy based on image segmentation. Biophysical Journal. 2004; **87**:2807-2817. DOI: 10.1529/biophysj.104.045492

[31] Jo JA, Fang Q, Papaioannou T, Marcu L. Fast model-free deconvolution of fluorescence decay for analysis of biological systems. Journal of Biomedical Optics. 2004;**9**:743-752. DOI: 10.1117/1.1752919

[32] Zhang Y, Chen Y, Li DD-U. Optimizing Laguerre expansion based deconvolution methods for analyzing bi-exponential fluorescence lifetime images. Optics Express. 2016;**24**:13894. DOI: 10.1364/OE.24.013894

[33] Sillen A, Engelborghs Y. The correct use of "average" fluorescence parameters. Photochemistry and Photobiology. 1998;**67**:475-486. DOI: 10.1111/j.1751-1097.1998.tb09082.x

[34] Fišerová E, Kubala M. Mean fluorescence lifetime and its error. Journal of Luminescence. 2012;**132**: 2059-2064. DOI: 10.1016/j.jlumin.2012.03.038

[35] Li Y, Natakorn S, Chen Y, Safar M, Cunningham M, Tian J, et al. Investigations on average fluorescence lifetimes for visualizing multi-exponential decays. Frontiers of Physics. 2020;**8**:576862. DOI: 10.3389/fphy.2020.576862

[36] Hauschild T, Jentschel M. Comparison of maximum likelihood estimation and chi-square statistics applied to counting experiments. Nuclear Instruments and Methods in Physics Research Section A: Accelerators, Spectrometers, Detectors and Associated Equipment. 2001;**457**: 384-401. DOI: 10.1016/S0168-9002(00)00756-7

[37] Maus M, Cotlet M, Hofkens J, Gensch T, De Schryver FC, Schaffer J, et al. An experimental comparison of the maximum likelihood estimation and nonlinear least-squares fluorescence lifetime analysis of single molecules. Analytical Chemistry. 2001;**73**: 2078-2086. DOI: 10.1021/ac000877g

[38] Turton DA, Reid GD, Beddard GS. Accurate analysis of fluorescence decays from single molecules in photon counting experiments. Analytical Chemistry. 2003;**75**:4182-4187. DOI: 10.1021/ac034325k

[39] Santra K, Zhan J, Song X, Smith EA, Vaswani N, Petrich JW. What is the best method to fit time-resolved data? A comparison of the residual minimization and the maximum likelihood techniques As applied to experimental time-correlated, single-photon counting data. The Journal of Physical Chemistry. B. 2016;**120**:2484-2490. DOI: 10.1021/acs.jpcb.6b00154

[40] Hall P, Selinger B. Better estimates of exponential decay parameters. The Journal of Physical Chemistry. 1981;**85**: 2941-2946. DOI: 10.1021/j150620a019

[41] Hall P, Selinger B. Better estimates of multiexponential decay parameters. Z Für Physics Chemistry. 1984;**141**:77-89. DOI: 10.1524/zpch.1984.141.1.077

[42] Köllner M, Wolfrum J. How many photons are necessary for fluorescence-lifetime measurements? Chemical Physics Letters. 1992;**200**:199-204. DOI: 10.1016/0009-2614(92)87068-Z

[43] Tellinghuisen J, Wilkerson CW. Bias and precision in the estimation of exponential decay parameters from sparse data. Analytical Chemistry. 1993; **65**:1240-1246. DOI: 10.1021/ac00057a022

[44] Kim J, Seok J. Statistical properties of amplitude and decay parameter estimators for fluorescence lifetime imaging. Optics Express. 2013;**21**:6061. DOI: 10.1364/OE.21.006061

[45] Laurence TA, Chromy BA. Efficient maximum likelihood estimator fitting of histograms. Nature Methods. 2010;**7**: 338-339. DOI: 10.1038/nmeth0510-338

[46] Chessel A, Waharte F, Salamero J, Kervrann C. A maximum likelihood method for lifetime estimation in photon counting-based fluorescence lifetime imaging microscopy. In: 21st European Signal Processing Conference (EUSIPCO 2013); 09–13 September 2013. Marrakech, Morocco: IEEE; 2014. pp. 1-5

[47] Marquardt DW. An algorithm for least-squares estimation of nonlinear parameters. Journal of the Society for Industrial and Applied Mathematics. 1963;**11**:431-441. DOI: 10.1137/0111030

[48] Johnson ML, Faunt LM. Parameter estimation by least-squares methods. Methods in Enzymology. 1992; **210**:1-37. DOI: 10.1016/0076-6879(92)10003-v

[49] Johnson ML. Why, when, and how biochemists should use least squares. Analytical Biochemistry. 1992;**206**: 215-225. DOI: 10.1016/0003-2697(92)90356-c

[50] Bajzer Ž, Therneau TM, Sharp JC, Prendergast FG. Maximum likelihood method for the analysis of time-resolved fluorescence decay curves. European Biophysics Journal. 1991;**20**:247-262. DOI: 10.1007/BF00450560

[51] Bajzer Ž, Prendergast FG. [10] maximum likelihood analysis of fluorescence data. Methods in Enzymology. Elsevier. 1992;**210**:200-237. DOI: 10.1016/0076-6879(92)10012-3

[52] Warren SC, Margineanu A, Alibhai D, Kelly DJ, Talbot C, Alexandrov Y, et al. Rapid global fitting of large fluorescence lifetime imaging microscopy datasets. Degtyar VE, editor. PLoS One. 2013;**8**:e70687. DOI: 10.1371/journal.pone.0070687

[53] Golub GH, LeVeque RJ. Extensions and uses of the variable projection Algorith for solving nonlinear least squares problems. In: Proceedings of the 1979 Numerical Analsysis and Computers Conference. Washington, DC: University of Washington; 1979. pp. 79-83. DOI: 10.5281/zenodo.852682. Available from: http://faculty.washington.edu/rjl/pubs/GolubLeVeque1979/index.htm

[54] Ware WR, Doemeny LJ, Nemzek TL. Deconvolution of fluorescence and phosphorescence decay curves. A Least-Squares Method. 1973;**77**:2038-2048. DOI: 10.1021/J100636A003

[55] Gafni A, Modlin RL, Brand L. Analysis of fluorescence decay curves by means of the Laplace transformation. Biophysical Journal. 1975;**15**:263-280. DOI: 10.1016/S0006-3495(75)85817-6

[56] O'Connor DV, Ware WR, Andre JC. Deconvolution of fluorescence decay curves. A critical comparison of techniques. The Journal of Physical Chemistry. 1979;**83**:1333-1343. DOI: 10.1021/j100473a019

[57] Apanasovich VV, Novikov EG. Deconvolution method for fluorescence decays. Optics Communication. 1990;**78**: 279-282. DOI: 10.1016/0030-4018(90) 90361-V

[58] Zhang Z, Grattan KTV, Hu Y, Palmer AW, Meggitt BT. Prony's method for exponential lifetime estimations in fluorescence-based thermometers. The Review of Scientific Instruments. 1996; **67**:2590-2594. DOI: 10.1063/1.1147219

[59] Fu CY, Ng BK, Razul SG. Fluorescence lifetime discrimination using expectation-maximization algorithm with joint deconvolution. Journal of Biomedical Optics. 2009;**14**: 064009. DOI: 10.1117/1.3258835

[60] Jo JA, Fang Q, Marcu L. Ultrafast method for the analysis of fluorescence lifetime imaging microscopy data based on the Laguerre expansion technique. IEEE Journal of Selected Topics in Quantum Electronics. 2005;**11**: 835-845. DOI: 10.1109/ JSTQE.2005.857685

[61] Liu J, Sun Y, Qi J, Marcu L. A novel method for fast and robust estimation of fluorescence decay dynamics using constrained least-squares deconvolution with Laguerre expansion. Physics in Medicine and Biology. 2012;**57**:843-865. DOI: 10.1088/0031-9155/57/4/843

[62] Dabir AS, Trivedi CA, Ryu Y, Pande P, Jo JA. Fully automated deconvolution method for on-line analysis of time-resolved fluorescence spectroscopy data based on an iterative Laguerre expansion technique. Journal of Biomedical Optics. 2009;**14**: 024030-024030. DOI: 10.1117/1.3103342

[63] Pande P, Jo JA. Automated analysis of fluorescence lifetime imaging microscopy (FLIM) data based on the Laguerre deconvolution method. IEEE Transactions on Biomedical Engineering. 2011;**58**:172-181. DOI: 10.1109/ tbme.2010.2084086

[64] Ballew RM, Demas JN. An error analysis of the rapid lifetime determination method for the evaluation of single exponential decays. Analytical Chemistry. 1989;**61**:30-33. DOI: 10.1021/ ac00176a007

[65] Chan SP, Fuller ZJ, Demas JN, DeGraff BA. Optimized gating scheme for rapid lifetime determinations of single-exponential luminescence lifetimes. Analytical Chemistry. 2001;**73**: 4486-4490. DOI: 10.1021/ac0102361

[66] Li DD-U, Ameer-Beg S, Arlt J, Tyndall D, Walker R, Matthews DR, et al. Time-domain fluorescence lifetime imaging techniques suitable for solid-state imaging sensor arrays. Sensors. 2012;**12**:5650-5669. DOI: 10.3390/ s120505650

[67] Collier BB, McShane MJ. Dynamic windowing algorithm for the fast and accurate determination of luminescence lifetimes. Analytical Chemistry. 2012;**84**: 4725-4731. DOI: 10.1021/ac300023q

[68] Sharman KK, Periasamy A, Ashworth H, Demas JN. Error analysis of the rapid lifetime determination method for double-exponential decays and new windowing schemes. Analytical Chemistry. 1999;**71**:947-952. DOI: 10.1021/ac981050d

[69] Silva SF, Domingues JP, Morgado AM. Can we use rapid lifetime

determination for fast, fluorescence lifetime based, metabolic imaging? Precision and accuracy of double-exponential decay measurements with low total counts. PLoS One. 2019;**14**: 1-20. DOI: 10.1371/journal.pone.0216894

[70] Li DD-U, Arlt J, Tyndall D, Walker R, Richardson J, Stoppa D, et al. Video-rate fluorescence lifetime imaging camera with CMOS single-photon avalanche diode arrays and high-speed imaging algorithm. Journal of Biomedical Optics. 2011;**16**:096012. DOI: 10.1117/1.3625288

[71] Li DD-U, Rae BR, Andrews R, Arlt J, Henderson R. Hardware implementation algorithm and error analysis of high-speed fluorescence lifetime sensing systems using center-of-mass method. Journal of Biomedical Optics. 2010;**15**: 017006. DOI: 10.1117/1.3309737

[72] Poland SP, Erdogan AT, Krstajić N, Levitt J, Devauges V, Walker RJ, et al. New high-speed Centre of mass method incorporating background subtraction for accurate determination of fluorescence lifetime. Optics Express. 2016;**24**:6899. DOI: 10.1364/OE.24.006899

[73] Li DD-U, Yu H, Chen Y. Fast bi-exponential fluorescence lifetime imaging analysis methods. Optics Letters. 2015;**40**:336. DOI: 10.1364/OL.40.000336

[74] Li DD-U, Bonnist E, Renshaw D, Henderson R. On-chip, time-correlated, fluorescence lifetime extraction algorithms and error analysis. Journal of the Optical Society of America. A. 2008; **25**:1190. DOI: 10.1364/JOSAA.25.001190

[75] Li DD-U, Walker R, Richardson J, Rae B, Buts A, Renshaw D, et al. Hardware implementation and calibration of background noise for an integration-based fluorescence lifetime sensing algorithm. Journal of the Optical Society of America. A. 2009;**26**:804. DOI: 10.1364/JOSAA.26.000804

[76] Digman MA, Caiolfa VR, Zamai M, Gratton E. The phasor approach to fluorescence lifetime imaging analysis. Biophysical Journal. 2008;**94**:L14-L16. DOI: 10.1529/biophysj.107.120154

[77] Fereidouni F, Esposito A, Blab GA, Gerritsen HC. A modified phasor approach for analyzing time-gated fluorescence lifetime images. Journal of Microscopy. 2011;**244**:248-258. DOI: 10.1111/j.1365-2818.2011.03533.x

[78] Ranjit S, Malacrida L, Jameson DM, Gratton E. Fit-free analysis of fluorescence lifetime imaging data using the phasor approach. Nature Protocols. 2018;**13**:1979-2004. DOI: 10.1038/s41596-018-0026-5

[79] Sorrells JE, Iyer RR, Yang L, Bower AJ, Spillman DR, Chaney EJ, et al. Real-time pixelwise phasor analysis for video-rate two-photon fluorescence lifetime imaging microscopy. Biomedical Optics Express. 2021;**12**:4003. DOI: 10.1364/BOE.424533

[80] Michalet X. Continuous and discrete phasor analysis of binned or time-gated periodic decays. AIP Advances. 2021;**11**: 035331-035331. DOI: 10.1063/5.0027834

[81] Vallmitjana A, Torrado B, Gratton E. Phasor-based image segmentation: Machine learning clustering techniques. Biomedical Optics Express. 2021;**12**: 3410-3422. DOI: 10.1364/boe.422766

[82] Vallmitjana A, Dvornikov A, Torrado B, Jameson DM, Ranjit S, Gratton E. Resolution of 4 components in the same pixel in FLIM images using the phasor approach. Methods

Applications Fluorescope IOP Publishing. 2020;**8**:035001. DOI: 10.1088/2050-6120/ab8570

[83] Vallmitjana A, Torrado B, Dvornikov A, Ranjit S, Gratton E. Blind resolution of lifetime components in individual pixels of fluorescence lifetime images using the phasor approach. The Journal of Physical Chemistry. B. 2020; **124**:10126-10137. DOI: 10.1021/acs.jpcb.0c06946

[84] Wu G, Nowotny T, Zhang Y, Yu H-Q, Li DD-U. Artificial neural network approaches for fluorescence lifetime imaging techniques. Optics Letters. 2016;**41**:2561. DOI: 10.1364/OL.41.002561

[85] Smith JT, Yao R, Sinsuebphon N, Rudkouskaya A, Un N, Mazurkiewicz J, et al. Fast fit-free analysis of fluorescence lifetime imaging via deep learning. Proceedings of the National Academy of Sciences. 2019;**116**: 24019-24030. DOI: 10.1073/pnas.1912707116

[86] Xiao D, Chen Y, Li DD-U. One-dimensional deep learning architecture for fast fluorescence lifetime imaging. IEEE Journal of Selected Topics in Quantum Electronics. 2021;**27**:7000210. DOI: 10.1109/JSTQE.2021.3049349

[87] Chen Y-I, Chang Y-J, Liao S-C, Nguyen TD, Yang J, Kuo Y-A, et al. Generative adversarial network enables rapid and robust fluorescence lifetime image analysis in live cells. Communications Biology. Springer US. 2022;**5**:18. DOI: 10.1038/s42003-021-02938-w

[88] Yao R, Ochoa M, Yan P, Intes X. Net-FLICS: Fast quantitative wide-field fluorescence lifetime imaging with compressed sensing – A deep learning approach. Light Science Application. Springer US. 2019;**8**:26. DOI: 10.1038/s41377-019-0138-x

[89] Xiao D, Zang Z, Xie W, Sapermsap N, Chen Y, Li DD-U. Spatial resolution improved fluorescence lifetime imaging via deep learning. Optics Express. 2022;**30**:11479. DOI: 10.1364/OE.451215

[90] Mannam V, Zhang Y, Yuan X, Hato T, Dagher PC, Nichols EL, et al. Convolutional neural network denoising in fluorescence lifetime imaging microscopy (FLIM). In: Periasamy A, So PT, König K, editors. Multiphoton Microscopy in the Biomedical Sciences XXI. Vol. 11648. Bellingham WA USA: SPIE; 2021:116481C. DOI: 10.1117/12.2578574

[91] Zang Z, Xiao D, Wang Q, Li Z, Xie W, Chen Y, et al. Fast analysis of time-domain fluorescence lifetime imaging via extreme learning machine. Sensors. 2022;**22**:3758. DOI: 10.3390/s2210375

[92] Héliot L, Leray A. Simple phasor-based deep neural network for fluorescence lifetime imaging microscopy. Science Reports Nature Publishing Group UK. 2021;**11**:23858. DOI: 10.1038/s41598-021-03060-x

[93] Xiao D, Zang Z, Sapermsap N, Wang Q, Xie W, Chen Y, et al. Dynamic fluorescence lifetime sensing with CMOS single-photon avalanche diode arrays and deep learning processors. Biomedical Optics Express. 2021;**12**: 3450. DOI: 10.1364/BOE.425663

[94] Li Y, Sapermsap N, Yu J, Tian J, Chen Y, Li DD-U. Histogram clustering for rapid time-domain fluorescence lifetime image analysis. Biomed. Optics Express. 2021;**12**:4293-4307. DOI: 10.1038/s41598-021-03060-x